The Theory of
Light Gravity

The Theory of
Light Gravity

Col Parkes

To order additional copies of this book, contact:
Xlibris Corporation
1-800-618-969
www.xlibris.com.au
Orders@xlibris.com.au
500490

CONTENTS

PREFACE

Why read this preface? Does the preface say anything that is not in the book? Readers often skip the preface and look through a book for bits that seem the most interesting or which provide the quickest answers. But why read the book at all?

Well don't be put off, just let me try and answer some of these questions. The preface is a bit like an introduction to a book and its author. In this case it should really help the reader feel at ease with an unknown author and a strange subject. And a preface is usually written by someone other than the author. However I could not find anyone who was as interested in this subject as I am and who was readily available, so I decided to write it myself. I hope it provides enough background to stimulate interest. And what about the book, well I wrote that myself, obviously, but I did ask some friends for comment and I have considered their responses and I thank them for that. It may seem a bit disorganised and repetitive and I have not followed a technical style that may have been appropriate for such a book. And while it is a scientific book I use non scientific language such as "perhaps", "hope" and "probably" and make suggestions based on how "I see things". But such issues shouldn't get in the way of a good idea and a new scientific theory. I am sure my new theory will eventually stand alone on its own merits. I strongly recommend that you read the preface and then the whole book. I am sure you will be rewarded.

The book describes my new theory of gravity. Perhaps the subject isn't very interesting so why did I study it and develop new ideas about it? What is there about gravity that is not already known you may ask? Most physicists certainly don't believe there are any problems with current gravity theory or at least none that urgently need fixing. But if you think gravity is a closed case you will be surprised. Almost every scientific theory has holes in it, mostly too small to worry about but sometimes too large to ignore. The current theory of gravity may seem complete but is not perfect and I was always trying to understand it more. Then I had a new idea, developed a new theory of gravity and wrote about it. But what is this radical new idea and what implications will it have on the gravity that keeps our feet on the ground. Well, it won't stop things like apples from falling, but it will surely explain why they do. My new theory solves many outstanding gravity problems and will definitely shake the scientific community. It will add to our knowledge and give new direction for science. Have I solved the mystery of the real cause of gravity by relating it to simple physical properties and forces, I believe so but will I convince anyone? Please read, I am sure you will be fascinated at what you find and hopefully you will also be convinced.

Why did I write this book? I certainly didn't write it for fortune as any income I might make from selling it in such a specialist area is very unlikely to cover the cost of production. And I am not seeking fame although any recognition would be nice. I am not an experienced writer and don't challenge the reader much, other that trying to get them to work out what I really mean. Welcome to my style, warts and all. But don't let my style or lack of depth put you off either. I hope it is not too boring for the experts or too complicated for others. Try and stay with it at least up until my new theory is described in some detail. Will this book suddenly explain everything you always wanted to know about gravity? Perhaps not, but it certainly should surprise you as you realise the significance of my new theory.

This book is a technical book but it is not a text book or a reference book. It contains lots of useful scientific information but only for background. It is definitely not fiction, although some readers may be wondering if it is all a hoax. Well let me say now that it is **definitely not a hoax**. This book is based on my desire to give a simple mechanical explanation for gravity. I like to describe things in simple laymen's terms without a need to refer to unexplainable forces at a distance and without a need to rely on complicated mathematics with no simple concept of what is really happening. While there is an element of myth busting about this book, I don't believe existing gravity theories are completely wrong, just incomplete. Of course Newton's theory of gravity was good enough to get man to the moon, but it has no real explanation of the cause of gravity. And while General Relativity (G.R.) provides a better mathematical answer, I have difficulty with space/time warping and such spin off ideas as the Big Bang and the expanding universe. This new theory, when accepted, will change our view of gravity. It will also have profound implications on current studies in many other areas of physics such as cosmology with its dark matter and dark energy and other dark ideas.

So who is Col Parkes? Where did I come from and what is my claim to fame if any? Who am I to review the great work of famous physicists and propose a bold new theory of gravity? I am not a well known scientist or a reincarnated genius and have never worked in a patent office. And I'm surely not in the league of brilliant men such as Newton and Einstein who have studied and written great theories of gravity before me. I am just an amateur or perhaps I should say an armchair physicist with a fascination for most aspects of physics and a drive to answer as many questions as I can, even in areas of physics already considered by many people, especially most physicists, to be fully answered. I am a pragmatist at heart and an engineer by profession and I don't like unsolved problems or impractical answers and need to relate everything to realistic solutions. So let me provide some more background to me and my interests in science and physics in particular.

My first exposure to science was through my fathers work. In fact I owe a lot to him for my interest in all aspects of physics. He was a signals officer in the Royal Navy during the Second World War and then moved to the Australian Navy. After a while he became a research assistant at the Research School of Physical Sciences at The Australian National University (ANU). His work started my interest in science and that university still holds a lot of fond memories for me. When I was a young boy in the early sixties, he would take me to work with him on school holidays. I loved to visit ANU but I'm not sure if site seeing was acceptable at the time but no one seemed to notice or mind. I am sure such visits are not acceptable these days and that is sad in a way but things were more relaxed back then. Anyway I was able to see some fascinating equipment and meet some interesting people. I saw complicated and perhaps dangerous equipment, and really enjoyed it. Perhaps it was all beyond my comprehension in those early days but some of it started to "rub off" and I was forever fascinated by science.

So with this background, I began my life long love of all things scientific. I always wanted to know how and why things worked. So after secondary school I became a science undergraduate at ANU, where it all started. I set out initially doing theoretical physics. It was a natural direction for me to take, especially after my experiences with my fathers work. But I didn't finish my science degree. Even though I started out this way I am not a theoretical physicist after all. What might have been? During my studies the Beetles music and pop culture were exploding on the scene. The anti war, peace and free-love movements were also gripping many university campuses around the world and many students were dropping out. I was caught up in it all so I dropped out too, stopped studying and set off to see the world. When reality finally returned, I returned to Australia to start a new life and take up my studies again. But this time it was in applied science and in particular electrical engineering. After graduating with a First Class Honours Degree I joined the telecommunications industry as a Professional Engineer and spent most of my career in engineering design and planning, moving into management at the latter stages of my career. But even then I was primarily a technical specialist. And now here I am continuing that role and writing on perhaps one of the most specialist subjects in physics and even proposing a complete new theory of gravity.

What are the benefits of this new theory? Does it solve important gravity problems and produce practical applications, such as an antigravity device? Well, unfortunately it doesn't at least not yet. I have thought about how I could patent my idea but unfortunately it seems to me that there is nothing to patent. But if anyone has any ideas I would be happy to listen and work with them. This theory and this book won't make me rich and I don't think it will make anyone else rich either. So, while it does not seem that my new theory will produce any immediate benefit, it is still a valuable contribution to science. Perhaps the idea will quickly be picked up by Hollywood for a new movie theme and I wish them well. Any publicity is great publicity they say. But when my new theory is finally accepted, it will make fundamental changes to gravity research. Considerable benefits may be achieved by redirecting resources into more fruitful avenues. New research associated with the new type of light should lead on to new technologies. It may not solve the energy crisis or reduce global warming but it may assist in these areas in due course.

So who will read this book? More to the point, who will believe it and take it seriously and perhaps even help to progress my work. I am sure there will be other amateur physicists who also have a fascination with gravity and light and have similar questions in their minds. I hope they will read it with interest. I also hope that it is read by some in the scientific community. Perhaps some specialist physicists who are into gravity and related subjects will be fascinated by the title and will try and read it through. Some may read it with a belief that perhaps there is something in this new theory. Others may read it with the intention of disproving it. Some may look at it just for fun while others may quickly just review it and dispose of it. Any reason would be acceptable to me. I just hope there are some with an open mind who read it and find enough value in this new theory to try and investigate it further. Some may even help to promote this bold new idea and for them I say thanks in anticipation. Read on and enjoy my ideas and hopefully come up with some yourself and be brave enough to publish them.

CHAPTER 1

A New Idea

Several scientific discoveries have really shaken the world and left their mark in many important ways. Some have been in geology, others in biology and more recently some in technology. But perhaps the most significant ones have been in physics. They have helped us to understand what makes the world go round, literally. Newton's Theories of force, mass and gravity; Maxwell's equations of light and Einstein's Theories of relativity immediately come to mind. Others include atomic structure, the periodic table, quantum light, nuclear activity and of course $E=m*c^2$. One thing they have in common is a bold and brave new idea which is often both severely challenged and strongly supported. These new theories become the focus of extensive research and when validated, are accepted by the scientific community. Great theories lead to important applications and often trigger even greater scientific discoveries. These bold and brave new theories become pillars of science and even pillars of society.

Then, occasionally a new idea comes along that challenges one of these pillars of science. Of course these days it is unlikely that any new idea will completely topple any of these foundations of modern science because they are so strongly entrenched. But there may still be room for ideas that give them a major shake up. A bold new idea could expose some weakness in existing theories but could also add further support in a positive way and give new direction to the physics involved. My new idea on gravity should do all of these. It identifies some weaknesses but also adds further support to most aspects of current gravity theory and provides a new basis for explaining the physics of gravity. It also adds a new dimension to the fundamental physic of light. I am sure it will lead to great benefits and go on to become part of the revised pillar of gravity and add further strength to our society. It will also create renewed interest and provide new direction in the exciting fields of cosmology and quantum atomic physics.

But first let me answer some basic questions about this book and myself. I am an amateur scientist with a fascination for physics. Whenever I hear or read about the work of the great scientists or about a new idea or discovery in the world of physics, I am enthralled. I like to tinker mentally with the physics behind such discoveries and I am always looking for a better solution to some well known problem. And yes to some extent I like the concept of myth busting. I like to examine new and old theories for weaknesses and inconsistencies. Even the best theories have some problems, so I consider alternative ideas and try to provide better answers to scientific puzzles. Then one day or perhaps once upon a time I had a great idea about gravity. I have always had doubts about the source of the force so to speak. The idea of undefined attractive force acting over a long distance has always puzzled me and I thought there had to be a more reasonable explanation. I believed the cause of attraction in gravity was more likely to be due to a simple mechanical force or a type of pressure effect. At the time I believed my idea was new and that no one else had thought about it. But it was so simple, how could it be unique? I soon discovered that it was not a new idea at all.

As I developed my idea further and did some basic analysis and carried out more research, I discovered that mechanical gravity theories were a reasonably well worn track with a long history. In particular, there is a long and well worn path of pressure or push gravity theories. My work and this book isn't the first on this topic. The idea has in fact raged since Newton's time and new variations seem to come and go from time to time. Some great physicists have dabbled in it, some to investigate its potential while others to try and disprove it once and for all. So I am not the first to study and propose a solution to the cause of gravity based on shadowing but I am sure I will be the last with this idea. My approach had an important difference and this convinced me I was onto something special and as you read on you will find out why. I really have added a new dimension to pressure gravity and have brought it back to life. I believe

I have discovered the real cause of gravity and this idea really excites me. While I have been driven by the excitement ever since it has taken a long time to get to this point but here I am at last. I have launched my great new idea into the physics community especially for gravity theorist to criticise, analyse and then hopefully support.

This book sets out to describe my new gravity theory. It starts by providing some general background to the subject of gravity and introduces some problems with current theories that led me to my new idea on gravity. It also tries to show how the existing theories of gravity are not incorrect, just incomplete and the reasons why I believe they are incomplete are explained. The book shows that gravity, or at least its cause, requires a better answer than is provided by existing theories, an answer that my new theory provides. But who is the book really for? Is it just for me to "let of steam" on a topic I am very passionate about or does it have a larger agenda? Well of course I did have some steam but the main reason for writing this book is because I wanted to share my new theory with everyone and I wanted to tell the scientific community about it. The question was: how best to do it. It is interesting to note that in earlier times, greats such as Newton and Darwin had published their work in books, mostly arranged by themselves or with the assistance of close colleagues. This may have been due to the fact that there weren't scientific societies with publications like those that exist today. But it may have just been seen as a better and quicker way to launch a challenging new idea.

But before I expose my new theory to criticism lets step back a bit and review some of the early ideas of physics and cosmology and use them to build a path toward current physics and in particular gravity theory. This may show how my new idea about gravity was triggered. Other areas of physics are included because it is difficult to clearly define the boundaries between general physics and the specifics of gravity. The connectivity between gravity and other areas of physics will become more apparent as the path of evolution of gravity theory continues. Of course at the end of the tunnel of physics and gravity is the so called unified theory of everything which seems to have become the holy grail of physicists. While my work on gravity does not provide a global answer to every question, I am sure it will make a useful contribution to many other areas of physics, including unifying theories.

THE SCIENTIFIC REVOLUTION

Science in general and the new area we now call physics came out of the dark ages, at least in the western world, in the renaissance period. This renaissance of civilisation awakened a new breed of natural philosophers and many became great scientists. They broke the hold of pure philosophy and introduced scientific research and analysis and made major breakthroughs in many fields of science. They moved away from dogmatic ideas about how things worked and challenged the non scientific doctrine of an earth centred universe. Science emerged as a real investigative profession and it also started to distance itself from some of the so called sciences such as alchemy and mysticism. Real questioning, analysis and scientific method were becoming the basis of this new scientific revolution. This started the never ending march of science that has continued until today and mostly helped the progress of mankind. But unfortunately in a few cases it has almost destroyed it. This book does not address any of the very early work on science by early civilisations such as the ancient western culture of the Greeks, eastern cultures such as the Chinese or middle eastern cultures. While such early civilisations may have studied what we call physics, including the question of gravity, there is little reliable documentation available today. This work was all but lost when these civilisations collapsed. Some early Greek philosophy was retained by Middle Eastern scholars who handed it on to become available for study by later western cultures. Such Greek philosophy became the source of most western knowledge and was basis for science even up to Newton's time.

Unfortunately Greek philosophy was not real science. While it was supposedly based on logical thinking by famous Greek philosophers they often just considered a problem and dreamed up a natural law about it. They did not try and prove these laws and did not do any guided experimentation and formal analysis. Before the renaissance these "natural laws" were taught in schools and universities for centuries as the only laws of science and they were not to be challenged. They were adopted and supported by the establishment, primarily the religious establishment of the time because they matched existing theology and supported the position of power the establishment had over society at that time. They were also used to some extent to justify the laws of society, especially religious based moral laws. But these philosophical ideas added little to the real development of science. At best they may have just slowed down any real scientific progress. At worst perhaps they set humanity back a long way when it comes to advances in civilisation due to scientific achievement. It would be a challenging but perhaps rather meaningless question to consider if scientific progress may have been faster without such Greek philosophy. Perhaps it was a necessary step along the path and helped to stabilise society and eventually to stimulate the collective brain of society into forming better questions on issues of natural philosophy. The correct formalisation of such questions was eventually a key step toward finding better answers to them.

One of the major natural laws of Greek philosophy was of course the relationship between the earth and the universe or should I say the heavens. It was based on a simple enough concept that human life was a very special life form and was unique in the whole universe. The earth was therefore considered a divine place which supported the view that the earth was at the centre of everything. But something was required to hold everything together. It was believed that all other celestial bodies were held in place around the fixed earth by fixed crystal spheres which rotated around the earth. Perhaps this was not a bad model for a relatively advanced but non scientific civilisation that was just starting the think about its environment, but it was too artificial and soon ran into problems. For a long time, too long in fact, this Greek model of the universe survived or at least was not successfully challenged. As more strange unexplained behaviours of the heavenly skies were observed that didn't fit the model these discrepancies were explained away based on new ad hoc laws or special adaptations of the model. For example, new observations of strange planetary movements were miraculously explained by the famous epicycles of the planets which were strange cycles within cycles. As more evidence built up against an earth centred model, the more the planets were made to dance in epicycles just to stay in place. Everything had to be kept in "orbit" rotating around a fixed earth. The model was convulsing itself to death but it just wouldn't die.

During the dark ages it seems that most free thought, especially in western cultures but perhaps also in many other cultures as well, was suppressed or at best discouraged. Any challenges to the existing hierarchy and their "scientific" models were considered an attack on the state and church and were quickly dismissed. An awakening may have started to occur earlier in some civilisations but it seems to have been quickly put down. Books and people were burnt to stop "wrong" ideas spreading. There was a gap of over a thousand years in any real scientific achievement which is a relatively long period in terms of recent human civilisation. This gap was eventually closed by the renaissance revolution. Around this time the Christian church in Europe was undergoing a reformation of its own. Challenges to the so called science of the time were starting to occur especially in the areas of astronomy. A developing society, a rapidly growing population, global travel, colonial expansion and emerging commercialism was causing increased demand, production and consumption of almost everything, especially non renewable energy, something that has now come back to haunt us. Bigger and better "tools of industry" were required and technologies such as steam power were emerging to drive them. Better solutions to existing and new problems were required and there was a need for continuous improvement, the forerunner to the quality and efficiency drives of today. There was new spirit of scientific enlightenment and a demand for better science and more scientists and. To meet this demand, education and research were expanding as never before. Modern society really started to embrace science and technology and to become its prime benefactor. Science was no longer studied just for its own sake; it started to earn its keep and place in society.

During the renaissance and the industrial revolution there was also an expansion of organisations dealing with topics such as masonry, architecture, engineering, chemistry, medicine and surgery. This led to the creation of art and masonry societies, craft guilds and many other technical and scientific organisations such as the Royal Society of England. They all contributed toward a push for improvement in the scientific process through the acceptance and use of more formal procedures for documenting, querying, approving and applying any new ideas and developments. They also started to control qualification and acknowledgment of participants in each of the fields. This growth in organisations and their supporting processes helped to formalise, document, distribute and apply modern science theory. Perhaps it also added a layer of cost not just in monetary terms but in terms of complexity and delay in promulgating new ideas. But without it science may not have had the solid foundations and not reached the heights it now has.

GRAVITY, CAUSE AND EFFECT

Our fascination with the concept of gravity must have been around for a very long time perhaps starting with mans early attempts to explain why things feel heavy or have weight and what mass or weight really mean. The word gravity comes from the Greek term gravis meaning weight. Perhaps the Greeks used a philosophical approach to answer the question about what keeps things on earth. But the real reason why things fall to earth or are "stuck" to the earth couldn't be resolved. Perhaps there were many theories, especially in early civilisations, as to what properties of matter made it fall. It was generally believed that things don't fall to earth at the same speed and heavier things fall faster. There was also the question about what keeps the universe together. However, as most questions, especially on cosmology, were abruptly "answered" by dogma, natural science was slow to progress. But as our observations and understanding of the vast cosmos grew, new questions arose. Why did only the planets seem to dance about the sky and was the earth a stationary central body or did the earth also move.

In the early renaissance days, the great cosmologist or early physicists of the time were busy making meticulous observations of the cosmos, and in particular of our own solar system with its dancing planets. Evidence against an earth centred universe or at least the solar system, was slowly gathering pace. Eventually the dogmatic answers could not be taken for granted any more. As technology improved and better observations and measurements were made, more questions arose and new scientist arose to answer them. Copernicus was perhaps one of the earliest to suggest a sun centred model fitted the data better. His work was expanded on by others, especially Kepler and the earth centred model became an unstoppable tide of science. Along with the acceptance of the new heliocentric model there was a need for new physics of mass, motion and force to keep it all together. A basis for this new physics was the work by Kepler into elliptical orbital mechanics which started to explain planetary motion in a more mathematical way. But a complete theory that governed the movement of the planets was still required. A number of proposals were made, some based on a kind of magnetic attraction while others used more mystical forces. Another scientist called Galileo was also working on the physics of moving bodies and his ideas became major inputs to gravity theory.

Galileo's work was extensive and he was perhaps one of the first scientists to correctly analyse force and movement. His studies and observations about the movement of rolling balls and dropped masses are well known, with or without the Tower of Pisa. His work was not as mathematically rigorous as later work but it provided an excellent start for others to follow. He was also one of the first to consider the question of relative movement which eventually evolved into the physics of relativity. He used a telescope to make startling cosmic observations such as moon creators and the rings of Jupiter. All evidence was pointing to the fact that other heavenly bodies were not perfect spheres but perhaps earth like formations that behaved according to some basic laws of motion. He was aware of the new helio-centric proposal and it matched his new discoveries so he supported it. His work was well known because of his high status in society. Others had similar views but not his position to promote them. But his popularity and his ideas nearly killed him during the inquisition.

The real awakening about what gravity was began during the late renaissance period. By now the helio-centric model was accepted science but the question of what kept it all together was still unanswered. The scientific revolution then slowly progressed and a more pragmatic and analytical approach started to be applied to explain how the new cosmic model worked. Many scientists were working on the problem and trying to solve the mystery we now understand as gravity. Serious physics and mathematics was being applied to determine the forces required to make it all stay in place or in orbit and to identify what may be creating those forces. Some of the scientist and some of the ideas they came up with are addressed in more detail in the chapter on the history of gravity but a few key points are covered here. Finally a new force was created to hold it all together. Gravity was about to be born, or at least a more scientific idea about what it was.

Then along came Newton. The name says it all but some more background on this great scientist and his work is given in the next chapter. He carried out extensive analysis of mass and movement. He even developed his form of the calculus (fluxions) to solve many of the problems associated with incremental movement and acceleration. He wanted to explain planetary orbits and comet trajectories, especially the one named after his close friend, Edmund Halley who encouraged him to carry out his studies and to publish his work and who helped finance him. By the late seventeenth century Newton had developed a theory of gravity so bold and powerful, that it stands today as perhaps the most significant single scientific achievement of all time. He presented his work in the now famous document simply called the "Principia". It is the centre piece of his brilliance and laid the foundations for modern physics, especially the physics of force, mass, movement and of course gravity.

But while his law of gravity was exceptional and solved most known gravity problems, it had some minor shortcomings. Of course many of these were not realised until much later when improvements in measurements showed up such anomalies. His theory had also not really solved the question of what causes gravity. He simply invoked a mysterious attractive action at a distance. This mystical aspect of his theory seemed to have caused him concern. But at least he had a plausible answer to what kept the planets in orbit and what gravity was all about and his law provided a comprehensive mathematical solution to how it worked or how to calculate its effect on bodies. He showed why the earth orbits the sun due to gravity and helped find new planets but his theory wasn't accurate enough to determine the correct orbit of Mercury.

Then another famous physicist emerged. His name was Einstein and perhaps just his name is also enough. His new theory of gravity started another revolution, not just because it solved outstanding planetary problems such as Mercury's perihelion advance but because it was so general and covered more than simple orbital mechanics. His new theory of gravity was part of a solution to the more general question of the relationship between force and acceleration, between mass and space/time. The new theory called the General Theory of Relativity (G.R.) involves complex ideas and even more complex mathematics and is beyond the scope of this section so I will leave this theory for more discussion in other sections of the book. But his ideas certainly changed our view of gravity and solved many problems that Newton's law couldn't. But just like Newton's law of gravity, this new theory also had problems.

SO WHAT IS GRAVITY?

The biggest difficulty in any serious review of a scientific subject like gravity is to define the problem or at least limit the scope of what is being studied. So the first thing to do is to better understand the question; "what is gravity?" Our understanding of gravity comes mostly from our experiences. Recent extra-terrestrial activities with manned spacecraft have given us more insight into gravity with astronauts and all their belongings floating in space. We have all seen such actions on TV and even seen man walking or bouncing on the moon in low gravity. These may have helped to build up our experiences of what we think gravity is all about but may not have helped clarify the question of what gravity really is. And of course we know that we can't always trust our experiences. So what else can be used to help clarify the question?

To better understand any ideas associated with gravity, perhaps we first need a more formal understanding of the components of gravity; that is of mass and force and even distance and time. However, I will not go into any formal scientific definitions of these properties or parameters at this stage and will proceed based on a general understanding that most people have about them. Problems this may lead to are addressed later.

In most situations, the simplest definition of something is usually the best one. However, terminology, as usual, often complicates things a bit. Terms like gravity, force of gravity, weight, mass and heaviness all have similar connotations. While not intending to complicate things anymore, a new term called gravitation is introduced. This is additional to the more common one simply called gravity. It covers a more general concept to try and simplify the definition. Perhaps it does or perhaps it doesn't, I'll let you be the judge.

The terms **gravitation** and **gravity** are often used interchangeably in common use, but in scientific use there is usually a special distinction between the two. Gravitation is used to describe the concept of "attraction" that objects with mass have on each other. On the other hand gravity more specifically refers to a force on a particular body that is supposed to be the result of such attraction (as in Newton's theory). That is, gravitation is the field or cause and gravity is the force or the effect. Now does that help?

So perhaps some better definitions of gravitation and gravity go something like this;

GRAVITATION: a natural phenomenon by which all objects with mass seem to attract each other.

Problems with this definition are it uses terms like "natural phenomenon" and "seem" which are incomplete or loose and is it proposes an incorrect attractive force.

GRAVITY: the force of "attraction" [again incorrect but let's proceed] between any (two or more) bodies that is directly related to the mass of each body and inversely to the square of the distance(s) between them.

This definition is simply a restatement of Newton's general law of gravity and as such is almost a mathematical formula. Inverted commas are used for the word attraction because while gravity appears to be an attractive force, and this is how Newton explained it, this is a misconception and is perhaps what set the physics of gravity back for a long time. There is a force but it is a push force, not an attractive force. It should be noted from another of Newton's laws that the force on one body from another is the same as the force exerted by that body on the first. The forces are reciprocal or are equal but in opposite directions and they apply to different bodies. The words "two or more" are in brackets because the force is on every part of each body down to atomic level (and even below?) and the total force between bodies is a summation of all the force components on each part of a body. So even a single body is still a composite of many bodies unless it is an atom and even then?

GRAVITY FOR THE MASSES

Modern physics describes gravitation using the General Theory of Relativity, but Newton's much simpler law of universal gravitation provides an excellent approximation in most cases. Newton's laws are used in secondary school physics and lower tertiary or undergraduate studies. The mathematics is very simple and is accurate enough for most earth bound and near space applications. It is also used in most scientific applications of a day to day nature. General Relativity is studied in higher undergraduate and post graduate physics and is used for relativistic situations and cosmic studies at extreme dimensions. Both theories are consistent for most day to day applications but Newton's law is used much more due to its mathematical simplicity. But there are some areas of orbital mechanics where Newtonian theory doesn't seem to fit the observations. Perhaps the most significant of these is the orbit of Mercury. It has a perihelion advance that can't be explained by Newton's law. However it is supposedly explained by G.R. which was one of Einstein's triumphs. Of course G.R. uses light speed for gravity and this alone supposedly explains the anomaly. This result was previously obtained by others who used light speed as the speed of gravity in connection with Newtonian mechanics.

The basic concept of gravity is well understood by most people and also quite comprehensively analysed by modern physics, but there are still some challenges and areas of interest. Not least is the unusual or some say mysterious concept of an "action at a distance". This aspect of gravity is not well explained by the two current theories, at least not in my mind and perhaps not in the minds of many others. This part of the gravity puzzle, action at a distance greatly challenged Newton and even he couldn't explain it and had to accept it as an unquestionable given. By contrast, Einstein tried to solve it using another concept that is not directly associated with any attractive force. Instead he created a new idea about space and time and said that gravitation is said to be due to changes in space/time geometry or something he called spacetime curvature. This curvature is apparently caused by mass or energy and is what causes objects to move (accelerate?) towards each other as they travel inertially (no force, no accelaration?) along "new straight lines" called geodesics. However, even given its apparent success, perhaps there aren't that many physicists, let alone ordinary people who can say without any doubt in their mind that they fully understand G.R. and "space/time curvature" and know how it happens.

Perhaps this raises the question; why did I really study gravity? Isn't it fully explained by all the theories developed by the greatest scientists or should I say physicist who ever lived! Doesn't Newton's law of gravity solve most of the problems or answer most of the questions we need to know about gravity? And then, doesn't Einstein's General Theory of Relativity fully explain all the mathematics as well as the reasons for and causes for gravity. And Einstein's theory reduces to Newton's equation of gravity for special non-relativistic situations? What is unknown about gravity and what is there to discover? What is there to achieve by studying it? Why bother? It would seem to most, if not all the scientific community and even the general public that any further study of gravity is futile and new theories are very unlikely to add any value. This is what may be called a subject of "low return". In fact, following requests for comment on my work from some university experts, a highly qualified Physicist, one of the few who acknowledged my letters, wrote back to me with the message "if it ain't broken then don't fix it". So therein lies part of the problem, most believe gravity or at least our existing gravity theory "isn't broken" and all the questions about gravity have been answered. There are no problems that remain unsolved and nothing needs fixing. These days the main research into gravity seems to be finding new examples that fine tune the proofs of the existing theories, or do they? Are there still any outstanding problems?

Now I for one believe there is always a need for further study! I think there is room for improvement in our understanding of all areas of science, even those that seem to be "fully understood". The study of gravity and in particular an explanation of the cause of gravity is one area of physics that I believe still needs to be examined. So, I set out to try and find a better theory of gravity or at least to try and find an answer to what seemed to me to be an unanswered question, "what really causes gravity?" Perhaps the real reason that I was fascinated by the study of gravity was not that I thought the existing solutions were wrong but they were not complete. I have been looking for an answer to the unexplained phenomenon of an action at a distance for a long time. I believe I have now found not just a better answer but the real answer in terms of a simple well known physical property. My new theory of gravity proposes a very simple physical explanation for the cause of gravity. It uses simple physical properties of matter, space, time and energy and does not require any type of strange unexplainable action at a distance. It may yet require complicated mathematics but not complex explanations. It doesn't throw out existing theories but fills their gaps. It fits in well with them and supplements them with a simple explanation of a real cause. It is simple and complete and easily understood by anyone with a reasonable understanding of physics. This work on gravity should also shed some interesting light on G.R. and the resulting theory of gravity based on it.

A NEW PARADIGM FOR AN OLD PARADOX

How can the old paradox of the cause of gravity be simply solved? Does it require a new paradigm in physics? Perhaps it does and perhaps I have found it. From my understanding of the background to current gravity theory, causality and the speed of light and I made my discovery. I had always believed that gravity was simply due to a causal force and realised that if the force acted at the speed of light then light must play a significant role. I also believed that gravity was not an attractive force but a type of push force and I also realised that if it was a push force, it must be due to a pressure difference caused by shadowing. The pressure was due to a kind of kinetic effect such as collisions from a type of high momentum background radiation or light. Of course this new light could not easily be detected apart from its gravity effect. There is also a delay in action just as there is in any other causal situation and this delay was due to the finite speed of gravity which is the speed of light. I then realised that the whole mystery of gravity was uncovered. As the title of my book suggests, my new theory is based on light pressure or perhaps more correctly lack of light pressure or light shadowing. Now while shadowing is not new, the light source behind it is very new and is indeed based on a new paradigm for physics. Together these two components completely solve the paradox of the cause of gravity. Again so simple and elegant and that is a key feature of a good theory.

So with this idea I was on my way to developing a new theory of gravity. It had to provide the same results as Newton's law for simple situations and produce the same type of results as Einstein's mathematics for relativistic situations. Newtons Mystery and Einstein's Enigma, terms used in the title, are solved and the more subjective issues such as Newtons Mistake and Einstein's Error are addressed. These somewhat emotional terms were used more to excite interest than criticize the greats. Who am I to know better than Newton or Einstein? At least their theories were relatively complete while my new theory still needs lots of work to finalise it and then to formally prove it. In particular what is still required is a complete mathematical model of gravity based on this new theory. In other words, the dynamic equivalent to Newton's static law of gravity using light speed or a type of matrix of differential equations is required to describe the mutual shadowing process. This is an area of considerable work and while I am happy with my limited progress so far, much more effort is required. Perhaps I may be able to obtain assistance from the broader scientific community, especially from gravity physicists, to progress this analysis. Any contributions toward finalising my theory and clarifying the ideas it is based on could be helpful. More on the opportunities for making contributions to this work later.

I know it is what is called a non standard theory or an alternative theory and yes it lacks rigor, but it does not completely negate existing theories; it just extends them in an elegant and scientifically sound way. This new theory doesn't change any general outcomes of gravity and any associated analysis such as why the planets orbit the sun but it shows why it works the way it does. It is also unlikely to cause any significant change to the application of gravity theory to find solutions to most everyday gravity situations. It won't really have any immediate impact on society or on any aspects of our daily lives, like some other recent scientific discoveries such as new medicines, but it is profound nonetheless. It definitely has the potential to make major changes to some areas of scientific research in gravity related fields. Many experts will undoubtedly say that this work is not scientific. I accept that the depth of analysis is not what is normally given in support of such a bold new idea in physics. But that should not prevent it from being considered in a scientific way. Some physicists may try and cut down my bold new theory by reference to some simple law of physics or more likely to some extremely complicated aspect of quantum theory. But I don't think this will happen, at least not as easily as some may wish. I can only hope that some may try to analyse it and perhaps help research relevant aspects of this new theory or even help to develop the necessary mathematics. I will certainly continue with my own research and analysis and try to work out how it may be confirmed.

When I had extended my initial ideas on the cause of gravity into a more comprehensive theory of gravity I wanted to launch it into the scientific world. This was easier said than done, especially for a relative newcomer to scientific research like me. While I may have been involved in some aspects of scientific research in my engineering days, I no longer have any contacts with people that could give me assistance in getting into any areas of physics let alone gravity. I tried several approaches to the scientific community, perhaps not as well directed and enthusiastic as they should have been or that may have been required, but all my efforts have proven fruitless. I should have realised that much more effort was required to get through the thick walls of academia. I wrote letters to some physicists in several universities but without any real response. It seems to me that the scientific community of highly qualified professionals is very busy and difficult to get any assistance from. This is especially the case in areas such as gravity which most believe has already been fully addressed by some of the greatest physicists that ever lived. What would a scientific nobody like me know? Someone who has a crazy new idea and has little if any background in this area surely has no chance. I also approached several of the more popular scientific publications and even trade journals or semi-scientific sections of the media that publish new ideas. However, I was unsuccessful in getting a hearing and they just referred me back to the "scientific community". Here we go again, round and around but not in ever decreasing circles.

So I decided to write about my bold idea. I am not a brilliant research physicist, just a humble engineer. I do have a scientific background in professional engineering so I am not a complete amateur in this field. However the fact that I am an engineer will probably make most physicists shake their heads in disbelief because engineers are usually too practical for real physics. The field of physics I am wading into, often called theoretical physics, is a strange world and is usually only for the so called "mad scientists", or those with little or no limit to their thinking. To them anything is possible. To engineers everything must be practical. What a contrast. How could I work on a theoretical idea from a position like this? Well the point is, it is not just theoretical, it is also practical. Some theoretical justification for my new idea is provided but it is mainly developed as a practical solution. It has a sound scientific basis but I have yet to complete the necessary analysis to develop the full mathematical theory. I just hope that this lack of detail and depth is not a major problem for advanced readers and they don't simply reject it on this basis. I also hope the new theory is not too complicated and can be understood by general readers. Let's hope it can be seen by both as a valuable new contribution to this fascinating area of physics.

Like some of the great scientist before me that had bold new ideas, I decided to publish my new theory directly. Not that I am any great scientist but my idea is definitely bold. And that is how I came to write this book. I'm not a very good writer. There is a lot of repetition, perhaps a few mistakes and even some obvious omissions. But I hope these don't distract from the sound basis of my work. I often provide too much detail on simple points and then give too little support for more complex ones. Sometimes I digress or wander off the main theme to make a minor point and then get lost before finding my way back to the story. I say story but this is not fiction, it is a real theory and I know one day it will be proven correct. I just hope I am still around when it is eventually accepted as the real answer to the cause of gravity.

The reader might ask whether my new theory provides any new information about gravity or forecast any useful new outcomes in gravity physics. Will there be any immediate new scientific developments as a result of my new theory? Does it show us how to control or prevent gravity and how to make an anti-gravity machine? Well at this stage I don't think there will be any big developments in the short term as a result of my new theory. In fact it generally supports what we already know about gravity. Orbit calculations won't change and our weight won't change either. This is unfortunate for me as the weight loss industry is a potential gold mine. This lack of any immediate financial gain from applying my new theory may limit press coverage. But there may be some interesting book and film opportunities and Hollywood is always good for a new idea. This new discovery of the real cause of gravity should, however, eventually lead to some significant changes in a number of major areas of scientific research, especially in the area of extreme gravity situations. In particular, it should change the direction of research in the fields of atomic or more precisely sub-atomic studies and also cosmic studies. Research into the Big Bang and Black Holes, may need to be reconsidered.

Of course the search for potential benefits from my new theory will undoubtedly be a fascinating area for future work. Perhaps it will produce beneficial outcomes and help us with some new and hopefully non destructive science endeavours. I just hope the first outcome is not a war machine like it was from nuclear research. One of the questions that I kept asking myself very early in my work was; what can I patent about my new idea or my new theory about gravity? The answer to that it seems is perhaps nothing. But I will continue to review all such possibilities. Also I wondered what indirect benefits I could expect to receive from my hew theory? Again, none it seems, apart from the satisfaction of knowing I have tried to contribute to the betterment of science and humanity. I am not really looking for any big rewards from this work, just the chance to make a contribution to science. Of course a Nobel Prize would be amazing, but this is perhaps a bit unrealistic, even if, or should I say when my new theory is proven correct. Perhaps I could be a candidate for the

IGNOBLE prize in physics. Of course I would also be very interested if any one is able to "prove" that it is all wrong. But I won't go as far as to say that if it is proven wrong then "someone up high" has made an error. But I will be very surprised because I strongly believe that this new theory will eventually be proven to be correct.

CHAPTER 2

Where Were We

This chapter introduces gravity, not so much what it is or what it does but where it has been or at least where our understanding of it comes from. To complete the picture, the history of gravity would not be complete without a broader view of physics. Physics as a science was created by those trying to find more logical and mathematical answers to questions about the real world. Questions about what stuff is made of, how things work and how everything is interconnected, challenge mankind. Answers to these questions had been proposed but as civilisation improved and rational thinking took hold these answers were often found wanting. Our current understanding of the world was gradually developed by great physicists studying subjects such as; mechanics; force, mass and acceleration; electromagnetism and light; and relativity. All of these are just as important as gravity and perhaps have also contributed to our understanding of it. While most of these areas may not have seemed to be related to gravity at the time and perhaps are still not, they played a major role in the complete story as my new theory will show.

The development of the theory of gravity from a purely mathematical point of view is based on only a few key ideas and discoveries. Fewer names seem to be associated with it than most other areas of physics. This is somewhat surprising given that it is such a fundamental part of our everyday existence. Perhaps this was because the physics of gravity was either so complex that few wanted to address it or so obvious that no one really thought more about it. Perhaps another factor was that the two major theories of gravity made such great leaps forward in their time and were practically so complete. They solved most known problems and no major work has been required since to address what are relatively minor shortcomings. Of course there are still outstanding problems that existing theories do not fully address. But most of these have only recently been identified with new technologies and more accurate measurements. These new found anomalies have recently triggered a renewed interest in gravity although they were not the real reason for my work on this new theory of gravity.

Of course Newton and his work was really the start of sound scientific gravity theory. Even today, Newton's mechanics and gravity theories are still the basis for most gravity science and analysis. His simple gravity formula still works well enough for almost all general earth bound gravity situations and most simple solar system orbital calculations. His approach and his equations are sound and straight forward. They are not rocket science but they have helped launch many successful rockets and will continue to do so. Then there are all the famous scientific achievements of Einstein, including his theories of relativity, especially the General Theory of Relativity (G.R.) which is the basis of current gravity theory and most new physics. It provides an answer to almost everything and is now the current scientific theory or standard model of gravity. It reduces to Newton's law of gravity for non relativistic situations but is applicable to most gravity applications at extremes of distance, mass, size and speed. G.R. has lasted for nearly 100 years and is still going strong. But it has some challenges and doesn't seem to fit all situations, at least not without "epicycles"

Some physicists have tried to extend Einstein's work in order to overcome some of the anomalous situations that have been identified. But there seems to have been little success so far. Will this approach ever lead to anything? This is an interesting question but according to my new theory this may not be very fruitful in future. There have also been some interesting attempts at developing quantum gravity theory but this also seems to have gone nowhere. And then there are the Theories of Everything (TOEs) and Grand Unifying Theories (GUTs) such as Super String Theory but so far these have just produced elegant mathematical models that do nothing for real physics. But who knows, they may help in biology. The solutions of G.R theory have also been widely used in other areas of physics such as the Big Bang theory, the expanding universe model and the theory of Black Holes. But some of these concepts may have serious problems according to this new gravity theory. So it seems there has been little if any progress of any significance in the area of gravity theory since Einstein's G.R. was accepted. No major new answers to outstanding questions have been forthcoming to date. Perhaps it is about time for a new theory.

So on with some history and great names in Physics. Sorry there no glossy pictures or photographs as most of these are copyright and many are just drawings anyway. And there are no graphics to show how some of the discoveries were made or how they work. I just hope the names and words are sufficient to create the same vivid images they have always created for me. If not there are many good books with more colour.

THE PHYSICS REVOLUTION

Physics literally came out of the dark ages, at least in the western world or in Europe, in the renaissance period. It emerged as a real science as it moved away from simple, pure philosophy and into framework of more sound and logical reasoning and proof. It also progressed beyond the somewhat mystical sciences such as alchemy and other "religious sciences". Philosophy had proposed an earth centred universe and there were other non scientific, primarily religious ideas on how things were created and worked. But now rational and logical thinking was coming of age. New ideas on what constituted real science, were emerging from European philosophers. All theories were being interconnected and questioned as never before. Kepler proposed a new sun centred model then Copernicus produced great work on planetary movements based on observations. This was expanded by others and became an unstoppable tide of scientific opinion and theory for an earth centred solar system. Copernicus published his work in a famous book which went on to become the centre of the new helio-centric model of cosmology and started the Copernican revolution. The unstoppable progress of science had started.

Along with the destruction of the ancient myth of an earth centred universe and the introduction of a more realistic heliocentric model of the solar system, there was a broader change in the physics of mass, motion and forces. While there were a number of earlier theories on gravity, major new work into formalising a theory of gravity and developing some more formal laws that governed the movement of the planets was required. While work may have been going on in other parts of the world, it seems as though the greatest strides in physics were being made in Europe from the 16th century onwards. The basis of the mathematical model was the work by Kepler on elliptical orbits. This was then extended by others into more realistic orbital mechanics that started to explain planetary motion in a more analytical way. At the same time, better laws governing the movement of bodies, at least on earth, were emerging. It was in this period in England that Newton emerged and developed his theories of force, mass and motion and also his theory of gravity. There were also many other great scientific and engineering discoveries that took place in England over the next few hundred years including Darwin on evolution and of course Maxwell on light.

The first section in this brief history lesson provides a list of names and events that led up to the great discoveries by Newton. The ideas of the early Greek philosophers Aristotle and Ptolemy are included as a starting point for this interesting journey of discovery. It then leads on through Newton's era and beyond and eventually on to the work of more recent physicists, especially Einstein. So let's look at some of the great names and ideas behind the development of our current understanding of gravity and the physics behind it.

ARISTOTLE (BC 112)

Aristotle seems to have played a major role in what was an early form of science called natural philosophy. He was a key player in the creation of the earth centred model of the universe which was based on Greek mythology with perfect shapes such as spheres and perfect circular orbits. Perhaps there were many others at the same time or before him with similar ideas but he seems to have been the one to give it formality. His philosophy was based on a simple observation of the night skies which shows the stars moving across the sky or around the earth in a circular way. This is easily seen in the night sky when the motion is speeded up. The other component was the mystical view that the heavens and all other heavenly bodies were perfect spheres. Hence all heavenly activity had to be spherical as it was the most perfect solid shape. And of course the earth was special because mankind was special or unique in the universe. Therefore it became natural in a simple sense to assume that humans must be at the centre of everything, including the universe and therefore the earth was at the centre of everything. This was the philosophy of the time and it became the basis of the earth centred model of the universe.

PTOLEMAEUS (AD 80-120)

Perhaps the most important of the earliest names in the history of cosmology and of planetary movements was Ptolemy. His name or at least his philosophy became the basis of early cosmic and hence early gravity theories. While perhaps he was not the first to propose an earth centred universe, he adopted the work of others such as Aristotle and set out to develop it more rigorously, if those words are appropriate to Greek mythology. He was by far the most influential in making the earth centred model popular for so long and it was named after him. He formalised and documented it so that it was the accepted science of cosmology and lasted for well over a thousand years after him. This was a long time in the short the history of modern human civilisation. But it was eventually overturned by real science from real scientists with open and questioning minds.

The cornerstones of the Ptolemaic model were the observations that all things seemed to rotate about the earth in the night sky, and that no earth movement could be detected or felt by man. It was therefore believed that there must be no earth movement and that the earth must be stationary and all other celestial bodies moved around our fixed earth. This immovable or fixed earth with all other bodies circling around it became the basis of early cosmology. These ideas also suited the religious and state dogma of the time and for a long period afterwards. These two mistakes of perception, if they can be called mistakes, were critical to the formulation and acceptance of the Ptolemy theory. They were based on simple human perceptions, shortcomings in human senses and on human egotism or at least on dogmatic beliefs. These false assumptions and the impressions they created were difficult to overcome. The first mistake or misperception, that if we can't feel or see something means it isn't there, has pervaded science and especially physics ever since. Human perception was a significant input, perhaps initially the only input, to the development of theories, until mathematics and technology came along.

Human perception or the lack of it has been used to justify many false ideas. It is perhaps behind some of the assumptions or misconceptions about current gravity theory.

The observation that all things seemed to simply rotate around the earth, was being challenged. Better observations were showing up more and more unusual patterns of planetary behaviour in the night skies. The planets seemed to do more than just rotate around the earth, in fact they sometimes seemed to stop or even go backwards. They became wanderers and the word Planet is derived from a Greek term meaning wanderer. This wandering of the known planets created a problem for the simple Ptolemy model. A solution was developed based on special movements called epicycles that kept them orbiting the earth. Epicycles were smaller inner orbits, and they were created to answer this apparent orbital anomaly each time it was found. The observation that we can't feel movement and the interpretation that therefore there is none, was more difficult to overcome. It still plagues scientific progress today. Fortunately both were eventually realised as errors and that finally removed the basis of the early cosmic model.

COPERNICUS (1473-1543)

He was one of the the new breed of scientific astronomers. He formulated a fairly complete *heliocentric* model of the solar system which removed the *Earth* from the center of the universe and put the sun in its place. He also proposed that the earth was not fixed but moved. While perhaps Copernicus was not the first to propose that the earth was not at the centre of everything, he was certainly one of the first to go public with the idea in a big way. This was at a time when printing and publishing were emerging and becoming a major vehicle in the promotion of new scientific ideas. But it was also at a time of great danger for anyone with ideas that challenged state and church. Because his ideas were so radical at this time and were counter to the existing dogma, they were considered to be a form of heresy. So he requested that his work not be widely circulated or published until after his death. He greatly feared the repercussions from the state for his ideas and they may have threatened his life and future. Another "scientist" called Bruno had recently been burned at the stake, apparently for expressing similar heretical views.

His book, De Revolutionibus Orbium Coelestium (On the Revolutions of the Celestial Spheres), published around the time of his death, started what became known as the Copernican Revolution. Perhaps it was the greatest scientific revolution of all times. The idea was founded on the assumption that the earth was not special and was not fixed but simply moved just like all the other bodies around the sun. His heliocentric theory was based on the idea that the apparent wandering of the planets, especially the higher orbiting ones of Jupiter and Saturn could only be explained if they rotated around the sun but further away than the earth which rotated closer to the sun. He also proposed that the other planets rotated around the sun but closer to it than the earth. He is often quoted as being behind the idea that the earth moves in many ways. He is also well known for the view that our sun and our complete planetary system is perhaps only one among many such systems in the universe. This is the basis of the so called Copernicus model and the now generally accepted Copernican cosmic model. His work started a wider acceptance of a helio-centric model and identified the need for a cosmic force to hold it all together.

BRAHE (1546-1601)

Brahe may not have played a direct role in the defeat of the Ptolemy model and the development of the new planetary model. In fact he supposedly supported a version of an earth centred model, albeit with some modifications such as some planets orbiting the sun. But his meticulous and well documented celestial observations which he made even without any special optical devices such as telescopes, eventually helped support the growing theory that the earth rotated around the sun. His work was well known to many astronomers and scientists, especially Kepler who became his assistant but only had limited access to this data. Brahe asked Kepler to work on his observations of Mars and these observations played a key role in further challenging the earth centred model.

When Brahe died, Kepler was in the right place at the right time and replaced him as court astronomer. After some hassles with the family, Kepler eventually obtained access to all Brahe's cosmic records. The detailed tables that Brahe had produced on the positions of all the planets, especially Mars, at various times of the year were invaluable. They helped Kepler arrive at his laws of planetary motion and helped him develop the idea of elliptical orbits. These two factors helped to confirm and correct the heliocentric model and they went on to be of major significance to Newtons work on gravity.

GALILEO (1564-1642)

Galileo is more widely recognised for the use of the telescope and his planetary observations and drawings. But he is also well known for his experiments with moving bodies and his theories of motion, in particular his proof that all bodies fall at the same rate. He is also considered by many scientists as the founder of the idea of relativity because of his observations of events on moving platforms. The real scientific genius of Galileo was in his use of observations and experiments to develop and support his scientific theories. He saw the moon up closer than ever before and realised that it was just like a piece of cratered rock. He also saw moons orbiting other planets, the phases of Venus and the rings of Saturn. This convinced him that all was not well with perfect spheres and earth centred model. He meticulously carried out and recorded the results of his experiments and was able to deduce correct inferences from his observations. He was aware of the Copernicus model and wrote a famous book comparing the Ptolemaic model and the new helio-centric one. The earth centred system was supported by a man called "simplico" (which can be interpreted today, just as then, to mean not very clever). The Copernican system was supported by a more intelligent sounding name. He had made his point but perhaps not too subtly for the church and establishment.

His revolutionary ideas and support for the Copernican view with his inference that only a simpleton would support an earth centred model, put him at odds with the church and in particular the inquisition. He was tried and threatened with death unless he withdrew his support for the Copernican theory. Eventually he agreed but during his recantation he supposedly muttered under his breath that the earth still moved in recognition of the Copernican model. He went under screaming, even if it was under his breath, and was placed under house arrest. This seemed to stop him from doing any further scientific research and he died soon afterwards. It was only very recently that the church eventually cleared his name.

While his work in these areas did not really achieve the mathematical level of others such as Kepler or Newton, it nevertheless played a key role in progressing science in the areas of mechanics and gravity. Galileo may also have been one of the earliest scientist to truly realise the significance and challenges of human observation in physics. He correctly realised that just because something can't be felt or observed by human senses, doesn't mean nothing is happening. For example in an enclosed space inertial movement may not be noticed. And we have all experienced the false sense of movement when stationary and something moves nearby like on a train or a in car. This is the reverse situation of not feeling actual movement but it still challenges our senses. Human senses are trained to be deceptive and to accommodate or reinterpret the environment, basically for our own protection. This understanding of the potential weakness of human senses has proven invaluable in real science, time and time again. Galileo was aware of it too.

KEPLER (1571-1630)

Kepler built on the extensive observations of Brahe. He was also aware of the work of his great contemporary Galileo and his observations with a telescope. Kepler also made many observations of the night skies with his new and improved telescope and was guided by his new planetary theories. He was perhaps one of the first to place the physics of astronomy on a firm mathematical footing. He used meticulous and detailed calculations based on simple mathematical formulae to support and justify his assumptions and his theories. This slow and painful work proved very useful in the future expansion of this science by others, especially Newton, who developed better mathematical techniques to make the calculations easier.

Kepler really put the new science of orbital mechanics on a sound scientific platform. His laws of planetary motion such as elliptical orbits and equal area for equal time law became key inputs to future theories of solar system motion and gravity which controlled it (although this was not understood at that time). Apparently he toyed with the idea of some force of attraction like magnetism, a phenomenon which was well known at the time, as being the basis for planetary attraction. He even had an idea of an inverse distance law for reducing the strength of the force of attraction, but it was linear and not inverse squared. However he was mostly known for the sound mathematical work he developed for supporting and modifying the heliocentric model.

DESCARTES (1596-1650)

While Descartes is not strongly recognised as playing a key role in planetary discoveries or the development of gravity theory, he is well known for his linear, three dimensional (Cartesian) coordinate system. He also wrote and published a paper on the physics of motion before Newton, also called Principia which had many of the ideas Newton later made famous. He had sound ideas on moving bodies and inertia, and these are considered by many to have been key inputs used by Newton in formulating his own mathematics of moving bodies and his theory of gravity.

LIEBNIZ (1646-1716)

Leibniz is placed on this list ahead of Newton even though most of his work came after Newton but he had some ideas before Newton. He is not as famous as Newton but he also had strong ideas on gravity. He was one of the first to firmly believe that gravity had to have a more mechanical explanation than simply a mysterious force of attraction at a distance. Leibniz was supposedly associated with the development of pressure gravity ideas and gave strong support for shadow based gravity theories. This work contained many of the ideas about forces between bodies that Newton eventually made famous by relating them to each other, developing them more rigorously and mathematically and presenting them in a more formal way. Newton was also able to promote them to others and this was a key to his success while other such as Leibniz didn't succeed.

Leibniz also developed his own form of the calculus at about the same time as Newtons Fluxion calculus was being developed. This started an epic copyright battle as to who really was the first to come up with the idea of the infinitesimal calculus with differentiation and integration functions? This contention over originality between the two famous mathematicians became a bitter battle that was never really fully resolved. Newton is generally given credit for developing his ideas first and without the knowledge of Leibniz's work. He was also definitely the first in applying it in the most productive way. But Leibniz form of calculus is more rigorous and became more acceptable and was the form that was eventually adopted by the scientific community, albeit after it was modified by others such as Cauchy. The calculus was essential to the development of the emerging new science of physics, especially for dynamic mechanics and gravity calculations among other problems. It continues to play a key role in physics and new developments still occur just as they have since it was first created.

NEWTON (1643-1727)

Of course no list of great physicists, especially in gravity and planetary motion, would be complete without reference to Newton. In fact Newtons work is still the centre piece of all our basic science in these areas. Not much needs to be said about the man because he is so well known. But I will provide a little general background to his great life and work. Sir Isaac Newton, to use his full title, was a famous English natural philosopher and theologian, who became a mathematician, astronomer, physicist and later an alchemist. He also became a politician for a short while and finally completed his long and distinguished carreer as finacial advisor to the crown of England. In this role he worked for the mint and helped to protect the English currency from many attempts at misuse and counterfeting. But it was his role as a physicist that he will forever be remembered.

Newton was undoubtedly a great scientific genius in his time and is still considered one of the greats today. He is generally considered by all scientists and most members of the public to be one of the most famous scientists in *human history*. His name will live on eternally especially in gravity, even with Einstein's work and also even after my new discovery is recognised. He also developed a new mathematical tool he called fluxions, a type of calculus, to help with the tedious calculations that his new physics demanded. Newton used his laws to derive the motion of objects on Earth and the orbits of known planets in the solar system. He finally proved mathematically that the helio-centric model was correct, and that all planetary orbits could be determined from his new laws of gravity and motion, proving that they are all governed by the same universal laws. Perhaps he started with the idea that gravity must be the same everywhere but now he had proof. This idea was expanded on by others when it was realised that perhaps all the laws of physics must apply everywhere and in every situation, an idea which eventually became the cornerstone of relativity.

He published his work in a book called "Philosophiæ Naturalis Principia Mathematica" (usually just called the Principia) in about 1687. This work is generally considered to be the most important book in the history of physics. It laid the foundation for physics for centuries to come. His three laws of motion and his law of gravity and their associated equations are of course important for almost every branch of physics today. Almost any student of any level of physics will know his name and also much about his work. They may not remember all his equations or know how to fully apply them but that is another issue. His theories have stood the test of time in prediction and application but like most theories they are just that, theories and they have some weaknesses. Newton's studies also covered another area of physics that is especially relevant to my new theory of gravity, the physics of light, This work is covered in the next chapter on light.

The scientific method of Newton was to develop a pragmatic theory based on known facts or observations, perhaps falling apples and other bodies near earth but especially planetary movements and derive a formula to fit. He assumed a simple law based on the inverse square of the distance between two bodies and the product of their masses (weight?). It is suggested he originally tried to use just the inverse of distance and he may have been swayed to try the inverse square law by his friend Hooke who proposed this idea, but it was Newton who eventually proved it was correct. His equation for the force of gravity between any two bodies is based on the product of the two masses and the inverse square of the distance between them or at least the distance between their centres of gravity (CG). He introduced a constant called the gravitational constant, G, and derived its value from observation and analysis, not from any basic theory or from any other constants of nature.

Newtons force of attraction "F", between two bodies of mass "m_1" and "m_2" which are distance "d" apart is given by the famous formula;

$$F = \frac{G * m_1 * m_2}{d^2}$$

This simple formula was a revolution in physics. It provided a solution to the known earth bound gravity and motion problems of as well planetary motion. It helped to confirm many cosmic observations such as comets and also make many predictions about our solar system and the universe. It works very well for most situations today and is still the basis of gravity studies up to lower tertiary level. One important point to make here is that Newton saw gravity as being causal. That is, at first a force occurred between any two bodies and this then caused them to try to come together. This direct causality aspect of Newton's gravity is significant when compared to other theories of gravity. But it is more significant when it is realised that he made no allowance for time or delay in his theory. There is no allowance for time or speed in his gravity equation, it is purely static. The fact that it worked at all was simply because the speed of gravity is so fast compared to the speed of the moving bodies he applied it to. But this is not always the case.

Newton's studies and theory did not determine what causes gravity. Perhaps he knew about earlier theories, some surprisingly based on pressure, but they didn't have wide support and it seems that he didn't support them either. This is even more surprising given that one of his close friends at Cambridge also formulated a fairly comprehensive pressure based gravity theory about the time of Newtons work. This mystery of an unknown attractive force of action at a distance may have caused Newton considerable concern and mental anguish, and yes even stress! It may have been what really forced him away from research and analytical studies that he enjoyed so much and drove him to spend considerable time and effort studying alchemy and religion. Perhaps he was looking for a divine solution to the cause of an action at a distance or perhaps he buried himself in his alchemy studies in an attempt to search for a gravity causing fluid? Some have said he may have believed that the cause of gravity was best left to mysticism or religion. He may have thought that it was the work of God so there was no need to consider it further. Whatever the reason was, it didn't detract from his fame and popularity and the great success of his work.

The other question that Newton's work didn't address was how quickly the force of gravity moved between bodies. Newton's model of relativity assumed there was an absolute frame of reference for space and time. He had thought about relativity in terms of the universal application of the laws of physics but his model did not really address time correctly. This may have led him to believe time and speed were not components of gravity. Perhaps he simply didn't see it as an issue and just assumed that gravity acted instantaneously, even over infinite distances. However the implications of the speed of gravity don't seem to have been very clearly addressed in his writings. Had he thought about it some more, perhaps he may have seen the light so to speak and realised that an instant force over infinite distance was not meaningful and a very high but finite speed had to apply. Again, perhaps, he would have been way ahead of his time with such an idea. Studies into very high speeds, especially the speed of light and other aspects of light such as Maxwells equations, which shed some more light on this whole question, were still centuries away. But while Newton may have initially assumed that gravity was instant, it must have become obvious to other physicists soon afterwards that this couldn't be so. The actual speed had to be finite and its value important, especially if any of the bodies in question were moving.

Much has been said about Newton and there are many good books on his life and work. One of the best I have seen is the Cambridge Companion series. However, I will leave it to the reader to follow these up for further background as required. It is worth saying that his theories and all his formulae for force, acceleration and gravity are sound and well proven and fully accepted by the scientific community. They are also fully consistent with my new theory of gravity to the extent possible and for most practical cases. Or perhaps I should put it the other way round; my new theory of gravity (TOLG) is generally consistent with Newton's theory in those areas that his theory is applicable. Much more will be said about his great work on gravity even after my new theory is accepted. And I feel certain that if Newton had been presented with this new idea as to the cause of gravity when he was developing his theory, he would have supported it. Perhaps he would even have been willing to include it in his great work. However, to fully appreciate this new theory, a great deal of knowledge about new aspects of physics and especially about light is required, all of which was completely unknown in Newtons time and for a long time afterwards.

HOOKE (1633-1701)

Hooke also had many ideas on gravity. In particular he had the idea of an inverse square law of attraction and it seems he made this suggestion to Newton. Hooke also did extensive research into light and made great use of the other new optical device of the time, the microscope. But it is unknown if any of his work in this area influenced Newton. Reference books on Newton discuss these aspects and especially his (sour?) relationship with Hooke, in a lot more detail so I will not say any more about his friends, associates or enemies here. However, I believe it is important to mention Hooke in the context of Newtons work, because it seems he in particular contributed more than is recognised.

FATIO (1664-1753)

Fatio or more completely, Nicolas Fatio de Duillier, was a Swiss mathematician who worked on zodiacal light, patented a method of making clock jewel bearings and did many other great things in science, especially in gravity. He knew many famous physicists and mathematicians and was also an acquaintance of Christiaan Huygens who later became famous for his work on light. He was a physicst by profession but perhaps more of an engineer by nature but he is not well known in either field. His name seems to have been lost over time. Of course Fatio is not famous for gravity and nowhere near as well known as Newton but he is known for his idea on "push" or "shadow" gravity. However, perhaps even here he is not as well known in this area as others who came after him, such as LeSage, who really only popularised Fatios original work.

The kinetic theory of gravity, or push or shadow gravity as it is sometimes known, seems to have been initially developed by Fatio in late 17th century. This was at about the same time of Newtons work on gravity. In fact Fatio also studied at Cambridge where he met Newton. Fatio was much younger than Newton and he was very impressed by Newton's work and supposedly had a very good (some even say a romantic) relationship with him while they were together at Cambridge. He would have learned much about gravity from Newton, probably well before he had any serious ideas on the subject himself. But it is not clear what direct influence Newton had on Fatios work or vice versa Again this is an interesting question but it seems the two became acquaintances well after Newton had developed his ideas on gravity. So perhaps the learning was more by Fatio.

More is discussed about his gravity theory in other sections of the book. His ideas are very interesting and in a way are very similar to mine. But as I said earlier, I did not come across him until well after I had developed the basis of my new theory. I was often of the view that my theory must have been considered by others before but I could not find any references in my initial quick research. It was only later after I put much more effort into investigating alternative gravity theories that I came across his work and the work of others that followed him in this area. When I found out about his ideas, I was very interested in his work and that of his successor Le Sage. It gave me some confidence and support for my own ideas.

LE SAGE (1724-1803)

Le Sage was another Swiss mathematician and is more popularly known than Fatio as the father of shadow or push gravity. He was often quoted as saying that he had many similar ideas about pressure gravity that were independent of and better than Fatio's work and that he had not simply copied the work of others. However, he was so impressed by Fatio's work that he amalgamated almost all of it with his own and only expanded on minor aspects of the theory. He subsequently formalised the theory of push or shadow gravity into a complete proposal but there is some suggestion that he added little of any value to the work of Fatio. But it seems that he helped to promote the theory to a wider audience and did at least make the concept more widely accepted as a real alternative gravity theory in his lifetime. Since then it has slowly faded into history until now.

OTHER GREAT NAMES IN EARLY PHYSICS

There are others that could be included in a review of great names in physics about this time. Some were associates of Newtons such as Edmund Halley, who had his own ideas on gravity. Others who may have known about Newton or even assisted him include his former professor at Cambridge and names like Boyle and Avogadro. These physicists were associated with improving our understanding of gases and matter at atomic levels. Their work was also relevant to the development of kinetic atomic theories which helped to generate interest in pressure gravity theories.

Such work was valuable for the development of physics but perhaps not so important to the initial development of gravity theory. More details on these and many more players can be found in the many detailed and interesting books on the history of science.

The next surge in activity that was relevant to gravity and related areas of physics was initiated by Faraday and Maxwell in the late 1800s. This was then followed by a string of discoveries and activity that led up to the work by Einstein in the early 1900s. From then onwards, apart from some key work on applications in areas such as cosmology and atomic research, things don't seem to have moved or changed significantly in the areas of gravity physics. Perhaps now is the time for a new development and I hope my ideas trigger that activity. But first let's review some of the key developments after Newton and up to the time of Einstein.

LAPLACE (1749-1827)

Following the work of Newton, which by now was widely accepted, the question about the speed of gravity was of concern to many physicists. Laplace came to the conclusion that it must be extremely fast and many orders of magnitude greater than the speed of light. Of course the finite but very fast speed of light was known about at that time but not well measured. More accurate measurements and Special Relativity was still many centuries away. But he offered no further explanation as to what the speed actually was or how the force of gravity was carried or propagated. His conclusion was very popular for a long time and it set the scene for a showdown on the real speed of gravity.

FARADAY (1836-1899)

Michael Faraday is included because he played a key role in formalising the study of electromagnetism which led to the comprehensive theory of light. Faraday carried out many experiments and proposed many new ideas, especially the ones about a connection between electric and magnetic fields. While his work was great and was widely recognised, he was unable to formalise it to the level that others eventually could. This was perhaps due to his limited education in formal mathematics. In a way, although I studied to higher tertiary level, I feel a bit like Faraday in not being able to fully formalise my new theory of gravity into a full mathematical model but I am still working on it.

MAXWELL (1855-1910)

The next great development in the history of physics is the work on electro magnetism by James Clerk Maxwell. His breakthrough in the theory of electromagnetism eventually proved light was electromagnetic in nature. Maxwell was a brilliant physicist. He came up with the real theory of light through the creation of his basic equations of electricity and magnetism. When these basic equations were solved they produced a wave solution which showed that electromagnetism was wave like in nature. In particular, these equations, now called Maxwells equations, had another surprising result. The speed was found to be the same as the current estimate of the speed of light. This resulted in the famous conclusion that light must be an E/M wave.

There were many other great names in physics that followed Maxwell in the study of the properties of light. Some addressed the question of its particle nature and properties. The most notable was of course Max Plank. His work is addressed later. There were others in the meantime who also contributed directly or indirectly to our understanding of gravity through the related question of relativity. Some of them are included in this review of great names of physics.

MACH (1838-1916)

I have included Ernest Mach in this review not because he played a key role in the development of any gravity theory but because of his strong ideas on relativity and in particular on inertia and momentum. He is perhaps most well known for the term Mach number which is about the speed of sound and not light. But he is also well known in scientific circles for his ideas on momentum and relativity and for a view on inertia named after him, simply called Mach's principle. His famous comments on the spinning water bucket experiment, an idea also mentioned by Newton, are often used to try and explain this principle but more on that later. One other thing Mach is known for is his serious doubt about atoms. Apparently he said if you can't see something it isn't there but we can see atoms or at least lots of them together.

PLANK (1858-1947)

Max Plank was a famous physicist at the turn of the 20th century. He is well known for solving the ultraviolet catastrophe. Plank solved this problem by realising the energy of light is related to its frequency or wavelength. His solution was a simple equation directly relating the two parameters. This equation together with the assumption that light was quantised into some type of "particles" was sensational and initially even Plank didn't believe it was correct. But it was subsequently proven to be correct and solved many outstanding problems and went on to become a cornerstone of physics. This idea was later used and extended by Einstein in his famous paper on the photoelectric effect, for which he was eventually awarded the Nobel Prize for physics. He didn't get it for his heavy relativity but for his light work.

EINSTEIN (1884-1955)

Just as for Newton, any work or book on gravity or even any area of physics must include the contributions of Einstein. His work on general relativity and gravity was published during the Great War. This followed his already famous work on special relativity. Progressing from the Special Theory of Relativity to a General Theory of Relativity was perhaps part of his aim to connect all the fields and forces of physics in various reference frames of movement, both inertial and accelerating. Perhaps he also saw the need to close the gaps in Newton's theory of gravity which were its cause and its speed. He must have wanted to determine the real cause of gravity and make allowance for speed in a more general theory of gravity or theory of everything.

Newton's gravity theory assumed an instant action over even infinite distances. This was recognised as requiring correction but few had addressed the issue. Some orbital calculations didn't quite seem to fit observations. Perhaps the most significant of these was the unusual orbit of Mercury. The reason for this is a relatively high orbital speed and the finite speed of gravity. Some physicists had "proven" that the speed of gravity was much faster than light speed to overcome aberration. Poincare, Minkowski and Sommerfeld, names also connected with relativity, were convinced the speed of gravity was no greater than light speed and in fact proposed that it was most likely carried at light speed and carried out calculations on the orbit of Mercury to try and prove it.

So what was the speed of gravity and how was Einstein to include a speed in his new theory? Einstein must have believed that gravity could not be faster than the speed of light. Gravity was like a physical force and was surely due to some kind of information transfer or field movement and it had to be consistent with Special Relativity. Gravity had to meet other speed requirements. It had to travel very fast over infinite distances. Einstein may then have realised that gravity must travel at exactly the speed of light. But if the speed of gravity was one of the driving forces behind Einstein's work, perhaps the other was the need to fill in the remaining gap which was "what causes gravity". If it travelled at light speed what did this mean about what its source was? What makes it travel at light speed if it is not light itself? How did Einstein handle this question?

He finally developed his General Theory of Relativity. It sounds so simple but it was a truly momentous achievement and perhaps one of the most significant in the complete history of physics. This theory was supposedly triggered by thought experiments of gravity situations but it is based on a mathematical approach to gravity. This new gravity theory assumed a form of background "field" which is somehow created and modified by mass and energy. The field is everywhere and somehow affects the properties of space and time. The idea seems to be very similar to my idea of virtual shadowing near a body and I wonder how this similarity will pan out. Any mass or energy curves or distorts space/time and this affects nearby bodies. The strength and directions of the curvature change according to movement and mass/energy. Any body creates curvature and also responds to curvature. It is mutually self correcting and accommodates movement of any body creating the field as well as any bodies moving through it.

Einstein used a new type of spatial geometry and associated mathematics for his theory of G.R. to allow for this complex mutual field interaction. It also accommodated accelerating frames of reference and produced conventional static gravity results. The mathematical way of describing this curved space was created by a mathematician called Riemann. The new mathematics was developed to explain non Euclidian space and is based on the use of Tensors which are multi-dimensional arrays of partial differential equations and metrics which are similar in a way to distance measurements of complex sloping lines. Riemann had already developed a stress/energy field tensor and another famous mathematician, Ricci, had developed a space moulding tensor called Ricci flow, which is related to topology and manifolds and was used for solving Poincare's conjecture. I won't explain this complex idea or tensors any more as I am still wrestling with them myself. But the challenging question is how Einstein went from his thought experiments and the equivalence principle to his solution based on this field of mathematics. Perhaps that is the genius of Einstein.

Einstein used the Riemann stress/energy tensor to represent curved fields due to mass/energy and the Ricci tensor to represent space time curvature (or is it the other way around?). At first he simply equated the two tensors but this didn't seem to work out and produced unacceptable results. He needed to add other factors or components for it to work correctly but these are beyond this review. Einstein finally came up with his new theory but the war initially prevented it from getting a wide audience, especially outside Germany. But eventually it was recognised as being of fundamental importance to physics and it became well known in scientific circles. Einstein became famous again.

His new "gravity formula" can be presented in a "simple" form such as;

$$R_{ab} = 8\pi \; G * T_{ab}$$

Or allowing for the cosmological constant Λ and other factors it becomes;

$$R_{ab} - (g^{ab} R / 2 + g^{ab} \Lambda) = 8\pi \; G * T_{ab}$$

Apologies if there are errors in my version of his formula but I did say it was complex and that I didn't understand it well. But I also said this was not a reference book and his formula is included just for illustration. And while this may seem a simple formula, it should be noted that the R and T are tensors or arrays of partial differential equations and other factors are metrics or other relations. In fact it is very complicated and even as Einstein himself said, extremely difficult to work with. It is interesting that Newtons gravity constant, **G,** is included in the equations and they also include the speed of light.

Many physicists started solving the equations. Some interesting outcomes were obtained but some problems were also discovered. Many experiments were also set up to test and verify the new theory. Perhaps the most famous was the solar eclipse observations to measure the bending of light which was used to verify the theory. Since then many experiments have supposedly proven G. R. to be correct. It is now generally accepted as the best answer to gravity. It produces basically the same result as Newtons law that gravity is directly proportional to the amount of matter (mass or energy) in each body and inversely proportional to the square of the distance between them. At light speed, G.R. also solves the Mercury orbit problem. But it still doesn't seem to fit perfectly with some areas of physics. It also makes predictions at the extremes of mass, energy, space and time which are currently the subject of considerable research.

There are some interesting questions that I have about G.R. The first is did light speed come automatically out of the creation of this new theory of curved spatial geometry or was it an input to the development of it? The second is why did the theory still need Newton's gravity constant in the equations and why didn't this also come automatically out of the appropriate aspects of the theory. Perhaps the answers to these questions are buried in the mathematics and I will need to improve my understanding of it to answer them fully for myself, if that is possible.

GRAVITY AFTER EINSTIEN

Following Einstein's famous theory there seems to have been little if any successful progress in the development of new or better gravity theories. Have all the problems been solved? Are current theories perfect? I don't believe so. There has been some work in areas like string theory which was going to solve everything but seems to have faltered. New theories such as MOND and Brans/Dicke and other reinterpretations of G.R. with scalar fields or some other complex variation have also arisen. But they don't seem to have produced any new startling results, only more complexity or in some cases perhaps just less applicability. Again I don't understand any of these new ideas well enough to be able to meaningfully comment on them but the wider scientific community is doing this, so far with no rewarding outcomes. Compared to the rearrangements of these G.R. deck chairs, my new gravity theory is completely different. It is simple and unique and did not start out as an extension of G.R. It may end up requiring similar complex mathematics but perhaps without a cosmological constant or even without Newton's gravity constant.

TOLG—NEW GRAVITY

My new theory of gravity, called TOLG which stands for Theory Of Light Gravity, was initially conceived some years ago and has been under development ever since. It is still not complete but I believe it is in a position that it can be launched for consideration. I have not put my name in the heading of this history item even though it is my theory and my book. Instead I will take a back seat and let the theory say it all and let others do the evaluation and write the history lesson. TOLG is still incomplete and it is not yet proven let alone history so perhaps that is another reason why it shouldn't be spelt out in this section. Tolg is explained in later chapters and there will be many more chapters on TOLG as it is developed, fully proven and accepted as the standard model. By then historians will have had time to absorb it and start writing it into the history of physics.

So is that the end of gravity. Will there be more gravity history even after my theory is proven or will any further work only be in the detail of this new theory? History will eventually tell all but I am sure there is still a lot to be discovered in this fascinating field of physics. I shouldn't be so presumptuous as to believe TOLG is the end of the gravity story. In fact I am sure it is just the start of a really interesting new phase of physics.

CHAPTER 3

Light It Up

"And then there was light" to quote or is it misquote a well known saying on the origins of things. But what is light? In simple scientific terms light is electromagnetic radiation. The sun is important as a light source, but there are many other sources of light in the cosmos. There are many forms of cosmic background radiation, some that can be directly detected and some that are perhaps less detectable. While some forms of electromagnetic radiation are obvious, there are others that are just as important but are less obvious. Perhaps light is involved in a lot more behind the scenes activities than we realise. This is especially true for my new theory of light gravity. The question of the relationship between light and gravity has fascinated me for a long time. It is what drove me to develop this new theory of gravity based on a new type of light. Without this light there would be no gravity, no earth and no life on earth!

The general term "light" is used throughout this book for simplicity to represent all types or perhaps more correctly all frequencies of E/M radiation and all types or sources of such radiation. The basic properties and "behaviour" of optical light, or light that we can readily observe, such as transmission, reflection, refraction, absorption are well known and reasonably well understood. This includes the wave like behaviour as well as the quantised particle like behaviour of light. These observed phenomena of optical light are known to be representative of all forms and all wavelengths of light or E/M radiation. Therefore what is discovered or observed to occur for one form of light is generally taken to apply to all forms of light across the entire E/M spectrum, given the different energies or wavelengths involved and the different properties of the materials involved as a function of wavelength.

I have placed this chapter on light ahead of the chapter explaining my new theory of gravity (TOLG) because light plays such a critical role in it. The type of light that is essential to my new theory of gravity is identified and some new ideas and key aspects of light that are relevant to gravity are analysed. In particular, light pressure which is critical to making this new theory of gravity work is explained. As a result of this new discovery, new challenges for the ongoing study of light and the extension of the theory of light will arise. Areas which are considered as clearly requiring more research and analysis are identified and will either be taken up by me or possibly by others in further studies of this fascinating subject. To use a well known phrase in a slightly different way, "we ain't seen the end of this (light studies) yet". I hope you will share that view, especially after reading this section of the book.

SOME LIGHT HISTORY

Initially light was considered as an emission of something from the eye that was sent to an object being seen. This emission was thought to bounce of an object and return to be captured by the eye and that is what enabled us to see. Not too far fetched when you think of RADAR. Eventually it was realised that the reverse was true and it was something that was received by the eye from the environment that gave us the sense of sight, not something the eye sent out. There was a lot of light from the sun or other sources already bouncing off objects. It was finally recognised that light must be some type of background radiation that is reflected by an object and is then detected by the eye. But what was the radiation and what was it made of?

The serious analysis of light was initiated by a few early scientists using simple technology. It was during this period, after thousands of years of darkness so to speak, that a theory of light started to be developed. Man "saw the light" and wanted to work out what it really was and how it did what it did. The development of new glass based optical technology began to be used in telescopes and then in microscopes and also prisms. Curved mirrors were also developed and used for telescopes for better cosmic observations. Following the development of the helio-centric planetary model there was a rolling movement of scientific study and analysis which led to the development of the theory of light we have today. While some of the basic properties of light such as reflection may have been known about long ago, the theory behind them was not well understood. It is interesting to note that the discovery path of light followed that of gravity in some respects. This relationship between light, cosmology and gravity continues today with these areas of science often boot-strapping each other to new heights of understanding. But surprisingly even with such a close relationship between light and gravity no direct connection had yet been identified until now.

The development of mathematical theory about the properties and behaviour of light started to gather pace. Properties of light such as colour and interference were not well understood until the work of great scientists such as Huygens, Newton and then Maxwell who supposedly made the first colour photograph of a piece of tartan. An extensive mathematical theory of wave properties of light was developed to explain interference and to explain diffraction. These new theories were also used to predict new types of behaviour, most of which was quickly found. It is interesting to note that many famous names involved with gravity also show up in a review of light. This may be a coincidence but perhaps not when it is realised how closely related these two fields of physics are.

THE TELESCOPE—GALLILEO

While Galileo did little directly for the study of the theory of light, his contribution was still significant. He didn't invent the telescope which was being primarily used for military applications, but he used it extensively for cosmic studies and helped to make it popular for scientific work. His optical observations and studies of the solar system helped with one of the greatest corrections of an incorrect "scientific theory" in history, the overthrow of the Ptolemy cosmic model. His work helped to establish optical based cosmic research as very useful and it continues as a major source of information in this exciting field of physics.

DIFRACTION AND THE SPECTRUM—NEWTON

The great physicists Sir Isaac Newton carried out many experiments and documented extensive research into the properties of light. He is more famous for his work on forces, mass, acceleration and gravity but he published a significant treatise on light in a famous document called *Opticks* which is still important today. Like most scientists, Newton was very interested in the unusual aspects of light that he observed in every day life or found with the help of simple optical devices. He made an early version of a reflecting telescope using curved mirrors which enabled a much shorter telescope to be made. This was a great advance over conventional refracting or lens based telescopes which were becoming too long for practical use. This new telescope also overcame fringe colouring that occurs in refracting telescopes. Newton was fascinated by colour and made a rainbow from white light using a prism. He discovered the spectrum and the decomposition of white light due to the process of diffraction and was one of the first to propose that white light was made up of colours which could be separated out and recombined. This theory of colour was not very popular at that time but was eventually proven to be correct. How close he was to discovering other spectral effects such as atomic line spectrum is anyone's guess. Another fascinating "what if".

One of Newton's most famous theories about light was that it was not continuos but quantised. He seemed convinced from his work with light that it was corpuscular or made of particles, although he did not appear to have any sound scientific basis for his idea. He may have been influenced by the kinetic theory of gases and the various particles theories of matter that were abound at the time. However, this was also at a time when light interference experiments were being conducted and presented in public forums as an early form of popular science. These showed that light seemed to behave in the same way as waves in water. Light was therefore popularly and scientifically accepted as having continuos wave like properties. Newton may have been influenced about quantum light by these interference results. Wave theory didn't seem consistent with a particles and it was considered impossible that light could be both quantised and wave like. If he had persevered he may have come up with some form of duality theory or quantum wave theory and perhaps the study of quantum science (or the 17th century equivalent of it) would have begun a lot earlier. He may have beaten Plank and Einstein to the theory of light quanta or photons as they are now called by over two hundred years but perhaps the necessary maths and physics weren't ready.

Again, while Newtons work on light may not have directly led to the development of current analytical light theory, he made considerable input to the science and was part of the revolution. It seems surprising that while the study of light was very popular in his time, he became more interested in the study of forces and gravity. Why did he concentrate his efforts on the more mechanical properties of nature than on light? Perhaps he couldn't see any real future in light theory and unfortunately he couldn't see any link between these two areas of physics. Also his colleague, Hooke was using a microscope, and looking at small objects under light. He produced some of the great early work on microscopic biology and helped to make this field of optical science popular. Perhaps Hooke's popularity in this field discouraged Newton to study light further.

LIGHT INTERFERENCE—HUYGENS, YOUNG, FRESNEL, OTHERS

Many experiments were being conducted into the phenomenon of light interference. Names such as Young with his double slit interference experiment and Fresnel with his lens interference experiments come to mind. A growing body of mathematical wave theory for light behaviour was also being developed by great names such as Huygens. Young's experiment is especially famous. It uses a single light source and two panels with holes in them with light shining through the holes in each panel in turn. The basic layout of the experiment is shown below. Light travels through the first hole which makes a single point source of what is called coherent light which then passes through the next panel effectively making two sources of coherent light. This light from the two sources then produces a pattern showing variations in light intensity just like waves in water. This experiment led to complete acceptance of wave theory as the complete explanation of all light behaviour. Young's experiment has been carried out using many types of light and even particles such as electrons, which also have wave like properties according to quantum theory. They always produce the same interference results.

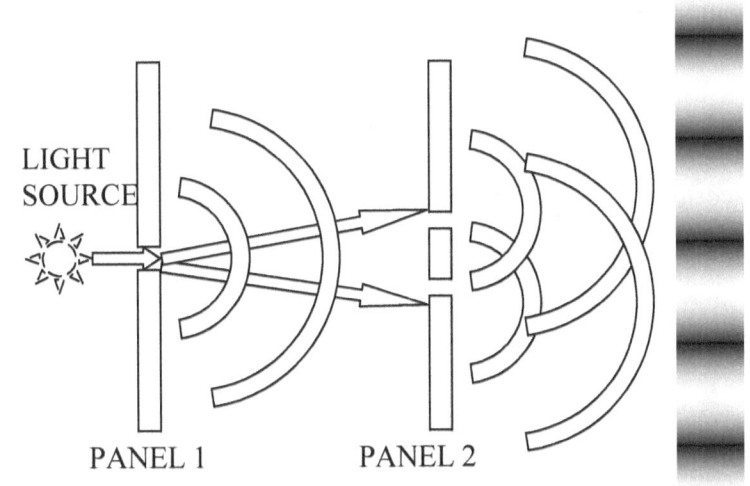

LIGHT
SOURCE

PANEL 1

PANEL 2

YOUNGS DOUBLE SLIT INTERFERENCE EXPERIMENT

What I find most challenging in this experiment is the explanation for dark regions in terms of light waves cancelling each other out. If quantised light theory is used (photons), that explanation is somewhat problematic. How can two or more photons cancel each other out just because they are coherent or out of phase or have some other different behaviour. The fact that the energy of the two photons somehow disappears seems very mysterious. This problem is supposedly addressed by saying each individual photon goes through both holes at once! In fact each photon supposedly goes in every direction, in a similar way to Feynman's infinite summation of multiple paths theory. When the two realisations of the same photon catch up with each other again they somehow add or subtract depending on the distance they have travelled. This idea seems to add more complexity and reminds me of the epicycles approach of those defending the old earth centred cosmic model. The mathematics seems impressive and it gets the results but is it meaningful or correct? Is there a better explanation?

ELECTROMAGNETIC LIGHT—MAXWELL (1831-1879)

Of course no discussion on light would be complete without reference to James Clerk Maxwell, one of the greatest scientists of the nineteenth century and of all time. He built on the work of other greats such as Michael Faraday, who also studied electromagnetism and who developed the basic concepts of field theory. However, Faradays lack of mathematical skills held him back while Maxwell used his great knowledge of mathematics to fully develop the theory of electromagnetism that we have today. Maxwell set the world of science on a new path with his famous equations showing the relationship between light and electromagnetism. These equations combined the separate E/M field theories into a unified E/M theory which eventually "proved" E/M "waves" exist and are the same as light. These equations known as "Maxwell's Equations" are well known to all science students and are addressed below just to remind us again.

MAXWELL'S EQUATIONS

Maxwell's equations express; how electric charge produces electric fields; how magnetic fields exist in dipole form without magnetic monopoles; how changing electric fields lead to magnetic fields and; how changing magnetic fields also produce electric fields. Some of these concepts and some of the equations may have been around before Maxwell but he uniquely re-derived them using Faraday's lines of force or field concepts. With help from other physicists and mathematicians such as Heaviside, Maxwell's original equations were reduced to four and put into the notation used today. These famous equations, still called Maxwell's equations, are shown below. I won't explain all the terminology and I apologise if they are not exact. They are only for illustration purposes.

DIV E = ρ/ε_0
(Charge density creates a field)
DIV B = 0
(No magnetic mono poles)
CURL E = -δB / δT
(Changing electric field creates a magnetic field)
CURL B = μ_0(J + $\varepsilon_0\delta$E / δT)
(Changing magnetic field creates an electric field)

These equations still send shivers down my spine and I love to see them but not to solve them. When the equations were eventually solved they produced electromagnetic waves. When this wave equation was examined and the speed of the waves was derived, it was found to be the same as the experimentally measured speed of light. This amazing discovery was perhaps one of the most important aspects of his work and made Maxwell and other realise that light is just another form of electromagnetic radiation.

The speed of the E/M waves is directly related to properties of nature, the magnetic permeability (υ) and electric permittivity (ε) of the media through which the E/M wave travels. The fact that this speed was a simple function of the permittivity and permeability of the media through which light travelled was just as surprising as its speed. Perhaps this important factor is still not fully recognised today.

The speed of E/M waves which is the speed of light c, is given by;

$$v = c = 1/\sqrt{(\varepsilon^2+\mu^2)} \text{ or}$$
$$c^2 = 1/(\varepsilon^2+\mu^2)$$

Where;

> ε is the electrical permittivity of the media and
> μ is the magnetic permeability of the media.

The Electric Permittivity, represented by the Greek letter ε, describes how an electric field affects a dielectric medium. It value is determined by the ability of matter or a medium to polarize or electrically allign in response to a field, and thereby reduce the fields strength inside the material. Thus, permittivity effectively indicates a material's ability to carry (or "permit") an electric field to penetrate it. In a capacitor for example, an increase in permittivity allows a given amount of charge to be stored with a smaller electric field and smaller voltage leading to an increase in capacitance. In general, permittivity is not a constant, as it can vary with the medium, humidity, temperature, the frequency of the field applied and other parameters.

Vacuum permittivity ε_0 (also called permittivity of free space or the electric constant) is given by;

$$\varepsilon_0 = 8.85 \times 10^{-12} \underline{F/m}$$

The Magnetic Permeability, represented by the Greek letter μ, is the degree of magnetization of a material in response to an applied magnetic field. It is measured in Henries per metre, or Newtons per Ampere squared. The constant μ_0 is known as the magnetic constant or the permeability of a vacuum, and has the exact or defined value of;

$$\mu_0 = 4\pi \times 10^{-7} \, N/A^2.$$

The next question to ask is; how do these parameters depend on any other properties of light? For example are they a function of the frequency, the intensity or energy or anything else? Of course it is now known that they are a function of wavelength and this is the reason for the dispersion of light and the rainbow effect. Lower frequency light (e.g. red light) is bent more in a media (e.g. glass) than higher frequency light (e.g. UV light) simply because it goes slower in the media than higher frequency light. The equation also implies that as the parameter become smaller then the speed of light increase and vice versa. If either becomes very large then the speed of light becomes slow. Perhaps there is a relation between these properties and Bose Einstein condensates where light supposedly can be slowed down considerably due to quantum effects.

It is also interesting that Maxwell's equations provided no direct indication of quantisation of E/M waves or the existence of photons. Why is this so? In fact Maxwell's equations and the wavelike properties derived from them were assumed at the time to be the final death knell of quantised light, but it rose again like the Phoenix. Can light quantisation and the properties of a photon be deduced from Maxwell's equations in any way? Or is there some other missing factor that needs to be added to these equations to arrive at this outcome. This is an extremely interesting topic for me and one that I have wrestled with since I first learned of Maxwells work and light quantisation or photons. I feel there is an opportunity here for some new understanding to be obtained. So how are Maxwells equations associated with quantum theory? This question has been addressed at some length and I am currently reviewing this area of the physics of light and the appropriate mathematical quantum theory with great interest.

There were many other great names in physics that followed Maxwell in the study of the properties of light. Some addressed the question of its particle nature and properties. The most notable was of course Max Plank. His work is addressed later. There were others in the meantime who also contributed directly or indirectly to our understanding of gravity through the related question of relativity. Some of them are included in this review of great names of physics.

THE ETHER—A LIGHT MEDIUM

One problem that challenged early scientist was the question of how light travelled. At the time it was sound science to believe that waves required a media to travel through. But what was the media that carried light waves? Eventually the aether, luminiferous ether or just the ether was created for light waves. It permeated through all matter and was also in and around us. Many physicists who supported the idea lead the search for this media but after much effort, none succeeded. Finally it was suggested that movement of the earth through the ether would lead to differences in the speed of light at different positions of the earth's orbital cycle. It was known that the earth moved at a considerable speed around the sun and this movement was in the opposite direction every six months. Surely such a large change in the speed of the earth would have a detectable change in the measured speed of light and this triggered a search to detect any such movement.

ETHER DRAG EXPERIMENT—MICHELSON/MORELY

Michelson and Morley, two great American scientists, wanted to measure the effects of the movement of the earth through the ether. They used a device called an interferometer which enabled interference patterns to be created between light that had travelled along different but orthogonal (right angled) paths. The device was built to be sensitive enough to detect very small light speed differences. I won't go into any details as they are well documented in many books on this great work but the experiment has been conducted many times since then in many different ways. Today lasers are used with very long paths and very accurate electronic detectors, not to detect speed changes but for other experiments. Of course, the results are well known; there was no change in the interference pattern regardless of how the machine was rotated or orientated or at different earth orbital positions over years of observations. The results were not due to poor equipment or bad measurement and have been duplicated many times with very accurate equipment. The results are always the same; no change in interference patterns is detected. This was immediately interpreted by most to prove that the ether didn't exist but some who supported the ether developed new explanations.

The results were interpreted in different ways by two famous physicists Lorenz and Einstein although Einstein supposedly said he wasn't aware of this experiment when he produced Special Relativity. Lorenz said the ether still existed and proposed that movement through it caused contraction of the measuring devices in the direction of movement and this contraction automatically accommodated for the speed change. He developed a theory of length contraction (Lorenz contraction) to explain it. On the other hand Einstein said light didn't need the ether and travelled at the same speed in any direction for any observer. This theory developed by Einstein, agreed with the Lorenz contraction concept that measuring devices in moving frames of reference contracted but not because of the ether but because of the constant speed of light. This interpretation by Einstein became his Special Theory of Relativity. It greatly challenged our understanding of light once again and eventually killed the ether theory or the need for a light carrying media, for good. But the theory of E/M was generating new waves itself.

RADIO WAVES—HERTZ, MARCONI

Perhaps one of the most famous scientists who took up the challenge to apply the results of the theory of E/M waves was Hertz. He managed to artificially create E/M waves or light, especially at lower frequencies using simple spark generators. His work confirmed that controlled transmission and reception of E/M waves, at least at lower frequencies, was possible. This was further extended and commercialised by Marconi who started the first practical radio communications service. This led to the creation of a whole new field of science and engineering associated with this new application of electromagnetic theory, called radio communications and it continues to develop even today. So far it has greatly improved our standard of living and helped create a better world through simpler and cheaper global communications. Some say it has made the world smaller and brought us all closer together and even may have helped prevent World War III. But is the internet overdoing it a bit?

LIGHT QUANTISATION—PLANK

Max Plank was interested in energy aspects of physics, especially the energy of light. He set out to solve the so called ultra violet (UV) catastrophe which was due to existing wave based light and energy theories. They created an uncontrollable runaway energy problem for higher and higher frequencies especially UV light. He came up with a theory which related the frequency of light to its energy and also proposed the concept of the quantisation of light energy. Plank showed that the energy of light is directly related to frequency. The higher the frequency the greater the energy light has but he also showed that light must be present in discrete packets of bundles of energy, an idea which was later supported and used by Einstein and which led on to quantum theory. His equation which relates frequency to energy is given below.

$$E = h * \upsilon$$

Where the energy is E, h is Planks constant and υ is frequency.

PHOTOELECTRIC EFFECT—EINSTEIN

The next major work on light was by Albert Einstein in the early 20th Century. While Einstein is most popularly known for his work on relativity he is also well known for his work on light. Any review of light or almost any physics subject would not be complete without reference to Einstein's work in the area. He was a good friend of Plank and was also very interested in the properties and behaviour of light. In two of his famous 1905 papers he put light under the microscope so to speak. His first work was on the photoelectric effect where he proposed that the only scientific explanation for some unusual photo-electric experimental results was that light was corpuscular or quantised in "packets". These were later called photons. He also proposed that Planks work could easily be extended to derive the energy of each packet (photon) according to its frequency. Quantum theory was born (not Max Borne, he came later).

Einstein's work on the photoelectric effect (for which he received the Nobel Prize) was instrumental in resurrecting the particle (quantum) theory of light. It was recognised however that light still somehow had wave properties so his ideas recreated the earlier Newton dilemma between the supposed continuous wave nature of light and the discrete modular nature of light. The reconciliation of these two properties was a concern to physicist at the time and still challenges physics students and experts today. But the idea of quantum theory is a mainstay of physics today. It has been used to explain many physical phenomena and will continue to do so. I wonder what impact my new theory will have on this age old dilemma. Positive I am sure.

SPECIAL RELATIVITY—EINSTEIN

Work on the photoelectric effect was then followed by his more famous work on the speed of light and its effects on measurement or vice versa. It was supposedly triggered by thought experiments based on riding a photon or was it a wave of light (an early form of surfing). This publication called "On Electro Dynamics of Moving Bodies" became known as The Special Theory of Relativity or just Special Relativity. In this work it seems as though he set out to study the dynamics of Maxwell's equations with moving observers and fields. He wanted to see how a moving observer saw changes in moving electromagnetic fields. In particular it seems he wanted to determine what the affects such movement would have on the measurement of the speed of light.

He arrived at the startling outcome that the speed of light is invariant to all (non light speed) inertial frames of reference. That is, the speed of light is the same for all observers travelling at any constant speed in any direction in what are called inertial frames of reference. Some say that he started with this postulate as a-priori knowledge but I am not an expert on how Special Relativity was "invented". Therefore, in a way, Einstein confirmed that the speed of light is absolute and not relative. Somehow this seems contrary to the whole idea of relativity where some physical properties such as time and distance (space) are relevant to the frame of reference they are measured in. However it is now universally accepted as the correct interpretation of light behaviour. This work was perhaps part of Einstein's aim to relate all the properties, fields and forces of nature together, a task he was unfortunately unable to complete. Nobody else has either.

LIGHT AFTER EINSTEIN

Other discoveries in light theory have helped to further our general understanding of the physics of light. Two that immediately come to mind are how the wavelength of light is related to electron activity or orbital jumps which led to the theory of line spectra and secondly the Radiometer used to observe and measure mechanical properties of light. The radiometer was first used in experiments in the 18th century to validate Maxwell's theory but the real importance of these results was noted much later. The radiometer and light pressure, which is critical to this new gravity theory, are discussed further below.

The connection between electron activity and line spectrum was a major discovery. It had been observed that elements gave off different wavelengths of light when heated. This was found to be due to discrete frequencies or line spectra. The property was also noted to occur in reverse when white light was shone through a gas and absorption lines were observed, These were the same wavelength but the exact opposite of the emission lines. Observing spectral activity provided a means of identifying elements but the reason for it was not clear. An explanation was eventually found and it was tied up with the planetary atomic model. The electron orbits were found to be related to electron wavelengths. This added to the understanding of electron energy levels and the development of atomic and quantum theory. Spectral theory was important to the development of many new areas of science, in particular, the observation of red shift measurements from some cosmic light sources which started the Big Bang theory of the origin of the universe.

There have been other developments into various aspects of the physics of light since these discoveries. Some experiments have helped confirm existing theories or have helped progress other areas of science but mostly only small steps have been made along a long path of hard work in a field of ever decreasing reward. Most recent work on light theory has not been worthy of inclusion in the Physics Book of Records. My new theory may change the direction of some of this research and open up a new field of light physics for new discoveries. At least let's hope so. But before going into any detail on the light behind TOLG let's look at light a bit more analytically.

KEY PROPERTIES OF LIGHT

There are several properties of light that really fascinate me. Perhaps one of most interesting is that speed is the same for all observers. It is also the maximum speed that information (or energy) can travel, according to Special Relativity. But how constant is the speed of light and does it change under certain situations, such as gravity? And there is the vexing question of what light really is. Is it a particle or a wave or both and if so what is the dual nature of light and how do particles behave like waves? And what about the source of light, where does light come from, what makes it? And there is the question about the mass of light particles and what this really means regarding the behaviour of light in a various situations such as in reflection. I will limit this discussion to a few key properties of light that I believe are worth special attention in particular due to their relevance to gravity and they are the;

- finite but very high (constant?) *speed* of light,
- schizophrenic nature or *wave/particle duality* of light,
- *photon properties* and activity,
- *energy/momentum transfer* of photons and

These properties are discussed in more detail below and while this work is by no means a thorough scientific analysis of light behaviour, it is based on well known general physics and is scientifically sound even if it is not mathematically rigorous.

THE SPEED OF LIGHT

One of the properties of light that challenged early mans thinking must have been how fast it travels? Perhaps the earliest view of the speed of light was that it was infinite or at least unmeasurable. But the idea of an infinite speed must have been seen to be inconsistent with other aspects of physics. But what was this speed, how could it be measured and was it always the same for all types of light or did it change under certain conditions? One of the earliest documented attempts to measure the speed of light was by Galileo. His method used light flashes between manned observation positions of measured separation. The time taken for a signal to travel between a sender and an observer was measured. He soon found that the speed of light was way beyond measurement using such relatively short distances and slow human response times.

Perhaps the first successful experiment to measure the speed of light used observations of planetary motion. Some cosmic events such as the time taken for one of Jupiter's moon phases increased when Jupiter was further from the earth compared to when it was closer. This difference was amazingly and correctly proposed to be due to the extra transit time of light. This fascinating observation led to an estimate of the speed of light of about *2.2 * 10^8 m/sec*. This was a surprisingly good result given the planetary distances at that time. At last the speed of light was captured so to speak.

This started the race to determine a more accurate measure and the focus shifted to using earth bound experiments. One of the earliest known scientists to have developed such a method was Facault, who used a mechanical device with large spinning wheels with slots. A light source was shone through the slots and the high speed was varied until a synchronous interference pattern was detected. The speed and dimensions of the wheel were used to estimate the speed of light and a figure of *2.8 * 10^8 m/sec* was determined. This was surprisingly accurate for a relatively crude mechanical device and the measurement systems available at that time.

Finally, experiments based again on rotating devices but using half mirrors to create longer paths, improved the accuracy of measurement down to about one percent. But while more accuracy could possibly be obtained with better experiments, there were fundamental problems with mechanical methods. There were limitations in the degree of accuracy that could be obtained of distance and time measurements and these set limits on the accuracy of measuring light speed this way. Today the speed of light is derived using lasers and atomic clocks. It is so accurate that it is now used for measuring distance, and not the other way around. The standard metre rod has been replaced by an atomic clock and the speed of light. Current methods using the latest high technology optical devices and solar system measurements have reduced the error down to 1 in100 million and this speed of light is now accepted as the SI standard. In most of my work in this book I have used an approximate light speed of 3 * 10E8 m/sec.

The next big step in determining the speed of light was theoretical. Maxwell and his famous equations did the trick. These showed the speed of E/M waves (light) was a function of other basic parameters. These parameters of the media in which the light travels or more correctly where the fields oscillate can be measured. From the two properties of the media or space through which the light is traveling, called ε and μ, the speed of E/M waves v or the speed of light c, is given by;

$$v = c = 1 / \sqrt{(\varepsilon^2 + \mu^2)}$$

The values of both of these parameters (or at least one), change for different media and this causes light bending. They also change as the frequency or the energy of the photons change. For high frequency light or higher energy photons, the values are lower than for low frequency light. This difference is quite small but it is sufficient to lead to a difference in the speed of light through a given media as a function of frequency and this is the basis for diffraction or the spectrum effect. While it may be very small for the "normal" range of frequencies of electron photon light, this may become an important factor for much higher frequency (higher energy) light and some media.

WAVE-PARTICLE DUALITY

The next mystery of light is the so called duality property. Light is a bit schizophrenic or perhaps to put it more politely suffers a bipolar disorder. It behaves in two ways or has two properties, that of a wave and that of a particle. There is an interesting parallel here with Fourier transforms between the time domain (particles) and the frequency domain (waves) but I will save that idea for future work. The behaviour of light shows a similarity to the behaviour of longitudinal water waves and follows the same mathematical theory of wave mechanics but that is where the similarity ends. Light "waves" are only lateral and have no longitudinal dimensions like other waves.

The particle properties of light are also based on experiments and theory such as that of Plank and Einstein. The work of these two great physicists eventually solved the issue of whether light was a wave or a particle, or did they? It seems that perhaps their theories added more confusion to the question about what light really is. And the debate over which is correct or which better explains the various behaviours of light has continued to rage since then. This apparent schizophrenic nature of light has puzzled scientist for a long time. I remember asking questions about this duality "problem" when I was a young physics undergraduate student. It was explained to me that a photon was something like a short wave packet or wavelet. It was perhaps a bit like a packet of waves as seen in surf waves by surfers. But I now know this is not a good explanation! It has longitudinal implications which are incorrect. So let's take a closer look at the photon side of light and try to get a better understanding of it.

THE PHOTON

The light "particle" was eventually given the name photon based on "photo" for light and "on" as a suffix for a particle, just like the electron. Each photon is a discrete and quantised "bundle" of E/M energy. Of course, there are many types or perhaps I should say "sizes" of photon. It depends on how they are created. For white light there are many wavelengths or colours involved while photons in monochromatic light such as a laser have (exactly?) the same amount of energy and the same wavelength. The measured energy of a photon also seems to depend on other factors such as the relative speed difference between source and detector or perhaps simply on distance travelled or age! The current theory is that the energy of photons is related to changes in energy in electromagnetic fields which create them. The amount of energy lost or gained by a field is the amount of energy transferred in the "creation" of a photon. Or in the reverse process it is the amount of energy gained in the "capture" of a photon. I have used inverted commas because I believe the full processes are not well understood.

Perhaps the most interesting aspect of light theory that seems to follow from relativity is the nature of photons. If light really is made up from little wave packets or wavelets how can these relate to photons. While wavelets fitted well with the conventional wave theory of light how do photons fit in? I am not sure how the wave/packet duality nature of light is explained today or how a photon is modelled but there are some key aspects that need to be considered. For light to be a longitudinal wave it must have depth or longitudinal dimensions. But if light is both a wave and a "particle" how can this be. A photon can't have any longitudinal dimensions at all and a simple analysis based on Special Relativity shows why. It states that anything travelling at the speed of light must either have infinite longitudinal dimensions hence infinite length or zero longitudinal dimensions. Now for a finite energy particle such as a photon to have infinite length would imply zero energy density at any one point or finite energy density would imply infinite energy and these are both meaningless. Therefore, light packets or photons which are travelling at light speed must have no dimensions in the longitudinal direction of travel. So a photon is only a two dimensional "particle". Therefore each photon can't be a longitudinal wave. It is purely a lateral wave which is observed to somehow have longitudinal wave behaviour.

This photon property of zero longitudinal dimensions was realised by other physicist about the time of Einstein. I am not the first to recognise this but it still seems to be misunderstood and it may require new interpretation in many discussions of light behaviour. My more realistic proposal for a photon is that is neither a wave nor a conventional particle. It proposes that light photons are discrete, modular, quantised or fixed size two dimensional discs of E/M energy. Each photon disc is made up of two orthogonal vectors, the electric or E field and the magnetic or M field which oscillate up and down or back and forth and mutually induce each other. This creates a spinning effect of the composite vector which is similar to the spinning field of a synchronous electric motor but I am showing my engineering background here. The vector product of these two fields gives the direction and magnitude of the light. Each bouncing lateral field creates a longitudinal wave effect when viewed from a lateral position and the composite scalar field creates a composite wave effect when viewed from any position as it travels in a helical spiral in the direction of travel. This model satisfies the quantum requirement for light and the special relativity requirement for zero length but what are its lateral dimensions and how does it contain mass?

Of course the photon disc can't have infinite lateral dimensions as this means infinite radial speed and it has unacceptable energy density implications. Both the E and M fields can't be ill defined and have no boundary or gradually thin out and. Each electric and magnetic field vectors and the photon disc must effectively be bounded. It must end completely at the boundary of the disc and have no energy beyond this boundary. It is therefore undetectable outside this so called capture area. The energy density across the disc may be variable but the size is fixed.

In fact photon "disc size" must be finite and related to wavelength ($d = \lambda$ or is it $d = \lambda/2\pi$?) and hence is inversely related to frequency and energy and hence "mass". It is interesting to note that this is the reverse of what may be expected. A bigger disc photon has less energy and hence even much lower energy density while a smaller disc has more energy and even greater energy density.

So each quantum of light or photon is like a spinning virtual flat disc with no thickness in the direction of travel. It spins at light speed orthogonal to the direction of travel and is observed to move at the speed of light in the direction of travel but it has no knowledge of movement if that is an appropriate term. The energy density of each photon disc is inversely related to the square of the wavelength and this is also important for the capture process. If a photon disc area is of a particular size (i.e. wavelength), then it can be captured by an object that has that "size". This is the basis for aerial sizes in radio technology which are wavelength related. It is also behind the process of photon capture by an electron in the appropriate orbit and hence of the required "size". If a photon is not the same size or wavelength as the capturing device, then it is not detected. It therefore passes straight through and is ignored. If the size of the photon and the capture device are the same or related (wavelengths) then the photon may be captured and an energy transfer made. Polarisation and other factors need to be taken into account in this process but these don't change the size requirement just the alignment. This capture and size situation must also have momentum and hence pressure implications but we still haven't worked out how mass fits into the picture.

This photon "flat disc" proposal is an interesting area for further study, especially in relation to the analysis of light interference. There may also be other areas of light behaviour that can be explained more easily using this disc model. Polarisation can be explained using this "rotating but fixed" disc model. The E field and M field are always orthogonal for any one photon and are effectively fixed in orientation. If the E and M fields in photons from a source are aligned so that the angle of the E field viewed from the direction of travel is the same for all photons then the light is polarised. Of course if there is no general alignment between all the photons and they are randomly aligned then the light is not polarised. The degree of polarisation depends on the degree of alignment. How do individual photons relate to each other apart from polarisation? Does light have an even more complex psychology to continue the analogy? Are there more sides to light than we know about? And do photons combine to behave in a wave like manner?

ELECTRON PHOTONS (EP)

Photons that we know about, work with in almost all areas of science and technology and have learnt to live with are created from electrons. I have called electron photons EP for short. Our understanding of EP light is that it is due to electron activity and electron energy level (shell?) jumps in atoms. There is a wide and continuous spectrum of EP light ranging from very low energy, low frequency, long wavelength light created by almost mechanical electron manipulation, up through optical light wavelengths and on up to very high frequency, "very short" wavelength EP such as X-rays and Gamma rays associated with high energy atomic activity. I put very short in inverted commas because this is a somewhat subjective measure. But what other sources of light or methods of making light are possible? What other particles can be "rubbed together" to make light? This is an area of special relevance to TOLG.

As described above, most of what we know about light comes from electron activity. But what other forms of light creation or capture could there be. Is light simply the result of the action of any charged particle? Now it is well known in quantum theory that all particles and even bodies have wave like behaviour. Even the sun has a wavelength associated with it. But is this purely a theoretical creation? Yes I can already hear the physicists asking; what is the difference? But I won't get into that debate just yet. Let's just continue with the conventional understanding of light and photons which are real.

So what other charged particles could be used to make E/M waves or light? It is well known that there are two charged particles in an atom, the electron and the proton. Both have exactly the same charge quantity which in itself is amazing, but they have different charge parities. The selection of a negative polarity for the electron and positive polarity for the proton is purely a convention and is not a real reason for any other difference. They do have significant differences in "size" or more scientifically in mass or energy content, but what other properties can be used to distinguish one from the other. It appears that there are no differences between the two that would make them behave differently, apart from the quark story which I will leave for later discussion. So the electron and the proton are both charged atomic particles that "exist" in the atom and therefore both should be capable of being involved in energy level jumps and hence in photon creation. End of story! Ooops there go some physicists.

PROTON PHOTONS (PP)

This is where a real jump in science is proposed. When I was developing my theory of gravity, light was central to it. Light pressure is what causes gravity through shadowing. But EP light didn't have the energy or mass and hence the momentum needed, it was too light so to speak. So I thought more about light in general and used lateral thinking to consider alternatives. Is light really only related to one of the two charged atomic particles? If so is there any obvious reason for this? Is it possible that perhaps the proton could behave in a similar way? Well I believe that the positively charged proton, can also create or capture light photons just as the negatively charged atomic particle, the electron does although the mechanisms are undefined yet. This is the basis of my proposal for light gravity. To me there is no obvious reason why such proton activation can't happen. The particular situations and types of atomic activity in which this process occurs are still unclear and are under review by me but I am sure will soon be identified.

This proposal of mine is a major break in conventional physics. I have not yet been able to find any other proposal along these lines in my (limited) research into atomic physics or into the theory of light. The current atomic (macro) model has the protons fixed in the nucleus but I believe this is incorrect and propose that they are in "orbit". My idea of a new atomic model is discussed more in the chapter on the atom. So now we have another type of light, proton light. I have called photons related to proton activity PP for proton photons, compared with EP for electron photons. Of course, it is assumed that PP and EP are both photons and both have light properties. If PP is associated with high energy proton activity which gives it high frequencies and hence short wavelengths, is it still covered by conventional light theory such as Maxwell's equations?

Maxwell's equations were originally based on electron related E/M field theory but they. Make no special distinction as to what causes the fields. So it can reasonably be assumed that Maxwell's equations apply to both EP and PP. But what about the properties such as permittivity and permeability, are they the same for both types of photons. If they are and if Maxwell's equations apply, then the speed of proton light should be the same as electron light. But this is a very challenging issue and requires much more study. It is also assumed that Planks theory applies to PP, just as it does to EP. Planks work was based on EP but seems to be a general concept which makes no distinction as to light source, energy levels or other such factors. Also Einstein's work on relativity and light speed was for EP light but it should apply to PP light. The only difference is in the level of energy involved and hence the wavelength, with PP light having much higher energies and hence much shorter wavelengths than EP light. Or in terms of my photon disc model, PP disc are very much smaller than EP discs and the capture process will need to take this size into account. But until some form of proton technology is developed to enable PP to be captured and measured and to test these assumptions of similarity or should I say equivalence, this is still an open question.

Wow!, this is new light indeed. Or to (mis) quote a famous space traveller,

> *Yes, there might be light out there . . . .*
> *. . . but not as we know it Jim, not as we know it.*

So the major difference between EP and PP is assumed to be in the energy levels and hence wavelengths or photon "disc size" as I call it. The other significant difference is that while we can see EP literally with our eyes and with electron based technology, we can't see PP but we can surely feel it in our weight. It is well known that the proton is several orders of magnitude "heavier" than the electron and must be orders of magnitude "smaller" than the electron. Therefore the energy density of PP must also be many orders of magnitude more than EP. Again not very scientific but let's stay with this general concept. The frequencies of PP photons are very high and wavelengths are very short and so they have very high energy and must therefore have very high momentum. From an order of magnitude analysis it is likely that PP have frequencies of order ~10^{25} Hz or greater, with corresponding wavelengths of ~10^{-15} M or shorter and hence energies of ~10^{-8} J or approx. 50 GeV or more. Of course these dimensions are first order guesses and much more work is required to support them. This simple analysis shows the relative energies and wavelengths that may be appropriate for EP and PP as set out below.

TYPE	MASS	"SIZE"	ENERGY	WAVELENGTH	ENERGY DENSITY
ELECTRON	1	~10000	1	~2000	1
PROTON	~2000	1 (??)	~2000	1	~1000000

So PP light is over 3 orders of magnitude "heavier" and up to 6 orders of magnitude "denser" than EP light. These increases have major implications on the capture or collision processes, the amount of momentum transferred in a collision and the impact of such momentum to create pressure. It is obvious that PP causes a much greater pressure than EP but is it enough for gravity? I believe it is.

It is well established that there is a certain level of background microwave radiation or EP radiation everywhere in space around earth and presumably everywhere in the universe. This is called Cosmic Microwave Background Radiation or CMBR for short. It was supposedly caused by the Big Bang, which I challenge, but I won't go into that here. It has an energy density which is equivalent to a temperature of a few ^{0}K. Also while CMBR was originally thought to be fully isotropic which means even in strength in every direction, better measurements have shown that it is "rippled" or variable in strength by direction. This is most likely due to the varying densities of CMBR sources in space, just like optical light is uneven. There is also some evidence that a form of directional "red shift" has been detected in CMBR. It has apparently been suggested that this is due to the earth's relative (absolute?) movement through the CMBR. (Shades of an infamous absolute frame of reference).

In my new theory of light gravity I have borrowed from conventional CMBR, and I have called PP background radiation Cosmic Gravity Background Radiation or CGBR for short. Now is CGBR isotropic or iso-cosmic? The term isotropic means equal in all directions but it has outward radiation connotations so I coined the term iso-cosmic for light coming inward to any point equally from every cosmic direction. It seems to fit better with cosmic sources of incoming light such as CGBR. There are some interesting questions about both CMBR and CGBR. If they are not iso-cosmic what is the amount of directional or time variation or the degree of rippling. CMBR has ripples so is CGBR also uneven or rippled? If so, then how much does it change? In local situations such as between the sun and the earth or the moon and the earth there will be variations in CGBR. This is what causes gravity due to differences in CGBR directional strength but are there other variations?

If background CGBR is not iso-cosmic this will have implications on gravity and the constant G. Even if it is only slightly uneven, how can such unevenness be detected. Any slight change in gravity that may result may be very small and unmeasurable in the short term but what are long term implications. Does the sun create PP even in small quantities? If it does then it will change the effects of gravity around it. If the amount of PP from the sun is only very small compared with general background CGBR, the effect will be small. But if it is a reasonable fraction of background CGBR then the reduction in gravity will be more significant because gravity pressure reductions from shadowing will be replaced by direct CGBR pressure from the sun. If so then this would imply that the suns mass is much greater than that predicted by Newtonian and G.R. gravity. Current estimates of the suns mass and volume produce a density of about ¼ of the earth's density. This density figure has been used to develop models of the suns structure, especially the internal makeup and likely fusion processes. If the mass is underestimated due to its CGBR then perhaps these models will need to be revisited. More work!

The big question is how can CGBR be detected and measured? Of course it is detected and "measured" by gravity but this is the conundrum. What other means of measurement could there be? There are no known methods available to do studies in this area but perhaps some cosmic experiments with gravity "situations" could be thought up to investigate this. Perhaps it can't be detected by using existing electron based technology which may not be capable of "seeing" proton radiation. So what type of technology is required? Can atomic reactions such as radioactivity and atomic nuclear decay be used in any way to try and "capture" PP? A possible source of PP or CCBR is atomic decay or activity at nucleon level. PP could be captured or released from proton "transitions" just like EP are from electron level atomic activity. Another way of creating PP could be from high energy particle accelerators like CERN. If so, then perhaps this may lead to an experiment for trying to capture them or at least observe them and measure their properties. Could a PP radiometer be set up in such an environment and be made to only detect PP by shielding it from EP? What a challenge.

ENERGY/MOMENTUM

This is not intended as a pun but one of the biggest mysteries about light is does it have weight, and if so, how much does it weigh? The question about the weight or should I say mass or is it the rest mass of a photon, has been and continues to be an interesting one. The other question about light that has challenged physicists for a long time is how does it transfer energy or momentum in a collision or capture process? These questions are key components of my new theory. From the work of Maxwell, Plank and Einstein, light is shown to have an equivalent mass and therefore momentum in its quantised form of photons. Each photon has a mass related to its energy which is directly related to its wavelength of frequency. The equations that provide these figures are given below.

Energy $\qquad E = h * v$

Where E is energy, **h** is Planks constant and **v** is frequency.

And also $\qquad E = m * c^2$ Where m is the "rest" mass and **c** is the speed of light.

So $\qquad m = (h * v) / c^2$

or $\qquad m = v * h / c^2$

Where h/c^2 is a constant of approximate value 7.7 10^{-50} which is a very small number.

Therefore the momentum of a photon, M_p, is given by;

$$M_p = m * c$$
$$= ((v * h) / c^2) * c$$
$$= (v * h) / c$$
$$\text{or } = v * (h / c)$$

Where *h/c* is another constant with an approximate value of 2.2 * 10^{-42} This is also a very small number but it applies to very high frequencies and to a very large number of photons to produce light pressure. The rate of CGBR is the great unknown. It may be possible to approximate it from CMBR data assuming it is a similar type of universal background radiation but PP sourced instead of EP sourced.

LIGHTS, PRESSURE, ACTION!

Or perhaps that is how the famous saying should go because light causes pressure which causes gravity which causes action. Maxwell was one of the first to come to the conclusion that light causes pressure. It came out of his work on electromagnetism. This was later proven in experiments by Russian Physicists Lebedev in 1900 and was confirmed shortly after by English physicists Nichols and Hull. Of course pressure from light as we know it, or from EP, is very low in intensity. But it can be used to move a suspended vane with certain surface properties that is supported in a glass chamber in a vacuum. This type of equipment is called a radiometer. The Nichols radiometer demonstrates light pressure, while the Crookes radiometer is driven by another phenomenon related to gas heating when the container is not fully evacuated.

Any body exposed to light will experience a force or pressure on it from that light. If the light is absorbed the pressure is the power flux density divided by the speed of light. If the light is reflected then the pressure is doubled. Of course if the light is absorbed and turned into some other form of energy which is then re-radiated, this may create a reactive pressure or a type of negative pressure effectively balancing the initial impact pressure. If the light is iso-cosmic there is no net force in any one direction. However, if the light is directional in strength then there is a net directional force. The sun creates directional light pressure on earth and every other planet. This is the basis of solar sails used to drive a spacecraft away from the sun. Light pressure from the sun is however very low. For example the flux density of the sun on earth produces a pressure, assuming the energy is fully absorbed, of about 5 10E-6 Pascals. If it is totally reflected it is twice this or about 10E-5 Pascals. The surface area of the earth is about 10E+14 sq. metres so this produces a net outward radial force on the earth by the sun of order 10E+5 Newtons or one hundred thousand Newtons. This is very small compared with the force of gravity on the earth from the sun so solar pressure has no significance in terms of gravity.

The theory behind radiation pressure can also be easily derived from momentum change of photons or from electromagnetic theory and thermodynamics. The Stefan-Boltzman law and constant relates energy and hence pressure to temperature in the formula;

$$E \sim \sigma \, T^4 / c$$

Where σ is Stefan-Boltzman constant, **T** absolute temp in K and **c** the speed of light

Pressure can also be derived from momentum. Light has mass and velocity and so it has momentum. The momentum of light or of each photon is readily calculated from simple analysis of photon energy. This gives an equivalent so called rest mass for a photon and together with its light speed gives a momentum. Of course the momentum of a photon is related to its energy and hence its frequency or inversely to its wavelength. The shorter the wavelength of light is the greater the frequency and hence the greater its energy and momentum. So very high energy, very short wavelength light creates greater pressure. For a complete elastic collision with a "hard surface", this transfer of momentum is turned into a pressure, P, where;

$$P = \rho * r * A$$
$$= v * (h / c) * r * A$$
$$= v * r * A * 2 * 10^{-42}$$

Where ρ is the momentum of a photon, **r** is the photon rate per unit of time and per unit of area, **A** is the area of the surface and v is the frequency of the photon.

There needs to be a factor which allows for the average angle of incidence and the angle of reflection of photons and this will reduce the pressure but not by an order of magnitude, only by a factor of less than 2. The other factor which impacts on light pressure from light striking a surface is related to the amount of reflection versus absorption. Again this involves at most a factor of two. For a typical mass of say 1 K.gm which has an area of about 100 sq. cm. or 0.01 sq m, the pressure required to support this mass must be 100 K.gm. per sq metre. This amount of pressure would be the CGBR pressure difference that causes the gravity effects on the body. This figure together with estimates of PP frequency will enable an estimate of CGBR density or rate necessary to create gravity pressure to be determined.

A LIGHT SUMMARY

The main outcome of this chapter is my proposal that there is a new type of light called PP or CGBR. I say *new* not because it has just happened but because I have recently discovered or at least proposed it. It has been there all the time and is all around us but like most light, we can't see it. But we can definitely feel it. Perhaps we don't realise what it does to us or means to us but it keeps our feet on the ground. This new form of light is just as important to our lives as electron light and especially optical wavelength light. While it works in a different way it has enabled our creation and is necessary to ensure our perhaps somewhat rough road to survival. All I have to do now is find it!

I hope that this chapter and these proposals will raise interest and not just eyebrows in the scientific community. Some of my proposals, especially new light sources, should trigger many reactions and hopefully some positive responses. And then there are my theories on the photon model which is sure to raise the temperature but hopefully the interest of physicist. I look forward to any comments from anywhere but especially and hopefully from the scientific community. I hope they are mostly positive but in any case constructive.

CHAPTER 4

Gravity by Light Pressure

This chapter presents the proposed new theory of gravity called TOLG. This theory is based on a new idea but is still generally consistent with current gravity theories including Newtonian gravity and Einstein's General Theory of Relativity. However, it offers a solution to the mystery of an unknown and unexplainable attractive force between bodies that seemed to challenge Newton and it also solves his speed problem. It also provides insight into the light speed enigma behind Einstein's General Relativity and provides a more plausible explanation than Einstein's complicated and abstract idea of a field that somehow changes space and time around a body.

So what causes gravity or the apparent "attractive action at a distance" between bodies? It seems to me that no one, including Newton and Einstein, has been able to satisfactorily work out this aspect apart from resorting to some mysterious action at a distance. Gravity theory has also been a challenge for unifying field proposals such as String Theory. It has not yet been addressed in a satisfactory way in any quantum work or in any proposed quantum gravity theory? Well my new theory of gravity based on light will certainly help solve these problems, but what is it all about?

My theory is that;

GRAVITY IS CAUSED BY LIGHT PRESSURE DIFFERENCES ACTING ON A BODY. THE LIGHT PRESSURE IS CAUSED BY A NEW TYPE OF HIGH FREQUENCY, HIGH MOMENTUM

LIGHT AND THE PRESSURE DIFFERENCES ARE DUE TO THE EFFECT OF MUTUAL SHADOWING BETWEEN BODIES. THAT IS, EACH BODY CREATES A SHADOW ON ANY OTHER BODY BY BLOCKING OUT SOME LIGHT AND THIS CREATES A NET IMBALANCE OF LIGHT PRESSURE ON EACH BODY WHICH FORCES BODIES TOGETHER—SO LET THERE BE LIGHT GRAVITY!

This new gravity theory, called the *Theory Of Light Gravity (TOLG)*, is based on the simple concept of a universal background light creating pressure on every body. Shadowing between bodies causes light pressure differences around a body and these light pressure differences push bodies together, hence the title "Light Gravity". This gravity causing light is a new form of background radiation from a new source of light. There is no longer any need for any mysterious attractive force pulling bodies together or a mysterious field that curves space and time. TOLG fits all basic gravity theory requirements and also has a few surprises that may change the direction of some areas of physics and cosmology research. And it is founded on simple physics. There is no smoke and no mirrors, well perhaps some mirrors but just to cause reflection of gravity light.

There is pressure from iso-cosmic CGBR light impacting on a body from all directions. This pressure is caused by the net momentum transfer that results from the impact of a new type of light photon. Momentum is conserved and transferred from each photon after a collision or more correctly an elastic reflection or deflection with the nucleus of an atom or hadrons (neutrons/protons). Bodies near each other cast shadows on each other and create pressure differences due to reductions in light strength. The net difference in pressure in all directions on a body, taking into account the reduction in CGBR strength from other bodies due to their mutual shadowing, produces a force on each body toward the other. Therefore, the force called gravity between any two bodies is simply caused by a pressure difference due to shadowing. So while the gravity force from TOLG is like an attractive force, it is in reality a push force. This is why it is often called push gravity.

This theory (TOLG) is so simple and elegant that it is surprising that no one seems to have proposed it before, at least to this level of detail. Some early work was carried out in this area some centuries ago by Fatio and others since then including Le Sage. Their proposal was based on pressure and shadowing, from some form of universal "fluid" as the cause of gravity. This is discussed more in the chapter on the history of gravity theory. However, there seems to have been little serious interest in this area since those early studies. Newton and Einstein must have been aware of pressure gravity theories but I am not sure if they were aware of the possibility of the pressure coming from some sort of light. I was unable to find any reference to it in their work during my research.

In "free space", the effects of CGBR shadowing on a body that is far from any other body, is very small. The light strength and hence pressure from all directions is almost equal assuming CGBR is iso-cosmic and there is no significant shadowing from other bodies. Of course in reality, there is always other matter somewhere out in space and all matter creates a shadow on all other matter. But if a body is at a great distance (light years) from all others it will effectively not "feel" gravity. However, if one body is "near" another body (even at large distances, just as it is in conventional gravity theory), then the two bodies (two masses) create a mutual shadow on each other which results in a net pressure to ***push them together***. This applies down to nuclear level and especially at proton level. This latter atomic action is a major area for further study and analysis in CGBR gravity theory. Does gravity due to CGBR light pressure difference only work on protons or does it also work on neutrons. What about electrons which have been "shown?" to have mass. Does CGBR create pressure on them?

In general any body made up of conventional matter deflects CGBR equally in all directions and hence effectively casts a kind of universal potential shadow all around. This shadow propagates at the speed of light just like the concept of a body changing space/time around it in General Relativity. If there is no other body nearby, then the shadow potential still exists and is created or sustained by reflected CGBR. If a body is moving, then the shadow has a latency effect (lag) due to the finite speed of light. The effect of such latency needs to be taken into account for gravity calculations on moving bodies. The "virtual shadow" is related to the ability of a bodies matter to reflect/deflect CGBR which is related to the amount of matter in a body which is of course its mass. Reflected/ deflected CGBR is retransmitted in all directions and becomes another part of iso-cosmic background CGBR. This momentum pressure on a body is iso-cosmic or completely symmetrical and is neutralised if there is no other body nearby to cause a mutual shadow (CGBR strength reduction). However, if two bodies interact by sharing shadows (i.e. each comes under the shadow of the other) then there is a net force (pressure) pushing them together which is *proportional to the mass of each body*.

The mass of a body is the (atomic) matter it contains. This is an absolute concept and is not simply its gravity "weight". Mass is related to the CGBR gravity by the density or the amount of matter (Nucleons or Hadrons??) it contains and which is used as a "deflection agent" to create a shadow. The real relationship between matter mass and gravity weight is also an interesting area of analysis in classical physics. It needs further analysis in view of this new gravity theory.

For bodies near each other, gravity attraction between them is simply the net CGBR pressure due to the mutual shadows of CGBR from all nearby bodies. The effects of mutual shadowing are directly related to the product of the mass of each body and it creates a force of gravity between each body which is proportional to the product of the two masses. So gravity is not caused by bodies being pulled together by some action at a distance but quite simply gravity is the net effect of CGBR pressure pushing them together, after allowing for their mutual shadows which are proportional to the mass of each body. The CGBR force between any two bodies, just like in conventional gravity theory, is of course identical. That is, a heavy body is pushed toward (attracted by in old terms??) a lighter one with the same force that the lighter one is pushed toward (attracted by??) the heavy one. This is because the effect of the mutual shadows is identical. That is, a large dense body creates a large (strong?) shadow but its effect on a smaller (less massive body) is reduced because of the size (mass) of the smaller body. Also the smaller bodies shadow is not as strong as that of the larger body but it falls on a bigger target. This type of reciprocal and equivalent action is fundamental to any theory of gravity and it applies so directly and simply with CGBR through the effect of mutual shadows. Hence this bold new gravity theory is equivalent in effect to conventional gravity theory.

Of course solid shadows are identical to the concept of the area of a shadow cast by a body in a sphere just like the concept of solid angle or area on a unit sphere, so to speak. It therefore follows that the *inverse square law* for a solid angle or area as a function of radial distance, applies to the strength of a shadow according to the distance from the body creating it. So that is the explanation for the application of the inverse square law for CGBR mutual shadowing and hence for shadow pressure strength and the resulting inverse square law of TOLG gravity force.

BASIC COMPONENTS OF TOLG

There are four major scientific concepts which form the basis of TOLG. These are all based on simple physics and apart from one area in the theory of light, they are not surprising new discoveries, But bringing them all together is the key to TOLG. The concepts are briefly addressed below and then further explained in more detail in later sections. The area of special interest in TOLG is the source of the light which creates the pressure. This is a major new concept and is addressed in more detail in the chapter on light. I could say "discovery" when referring to this new form of light but today it is only a proposal and I have yet to discover it or more correctly to confirm its existence. That is going to be the basis of ongoing research by me and I am sure by many others.

The four components of TOLG are;

1. A new form of background radiation (E/M radiation or light photons) which has very high energy and hence high momentum. The energy and hence frequency of this gravity light is much higher than electron based light. This light (E/M radiation) is called ISO-Cosmic Gravity Background Radiation (ICGBR) or just CGBR for short in a similar way to CMBR. I have coined the term Iso-Cosmic for radiation that is always coming inward to every point or body from all directions in equal strength. The term Iso-Cosmic is similar to the term Iso-Tropic which is more typically used for radiation that is going outward in equal strength in all directions from a source. This difference in terms is used to explain the inward (as against outward) properties of this special gravity light which is so important to TOLG.

2. Background radiation or light creating a pressure on a body when it impacts on that body. Pressure which results from the change in momentum of light (photons) bouncing of a body, is a well known and simple concept. The average pressure is the accumulation over

time of the effect of a large quantity of photons being deflected and hence experiencing a momentum change. To conserve momentum, this momentum change is transferred to the deflecting body which causes a pressure on that body. Some types of radiometer show the effect of light pressure, by the movement of a balanced vane. The net light pressure on a body and its direction is central to TOLG. The pressure on a body from CGBR is proportional to the amount of matter in the body. Hence the force of gravity is proportional to the amount of matter in each body which of course are the (gravity) masses of the bodies.

3. Mutual shadowing between any two bodies. That is, when two bodies are close together (or in fact any distance apart!), they each form a shadow from background radiation (CGBR) on the other. This shadowing reduces the amount of radiation pressure (CGBR) coming from the direction of the other body. This causes an imbalance of radiation pressure on each body which effectively pushes the bodies together, hence a type of force of attraction or gravity. Mutual shadow strength is dependant on the amount of CGBR deflecting matter in a body as well as deflection mechanisms.

4. The strength of mutual shadowing is inversely proportional to the square of the distance between the two bodies. The inverse square law is simply related to the concept of solid angle which shows the relationship between the area of a shadow cast by a body at a distant point and the distance between that point and the body casting the shadow. The solid angle (area) is inversely proportional to the square of the distance between the body creating the shadow and the body receiving it and of course vice versa. This produces the inverse square law effect on pressure difference and hence on the force in TOLG which is central to any gravity theory.

These concepts, which are simple and elegant, are explained in more detail below and in the other chapters. They combine to form the basis of TOLG.

BACKGROUND RADIATION

This is not a new concept. Light from the cosmos has been observed since time began so to speak. A different type of background radiation,

CMBR, was recently discovered and has been studied at length since it was first detected and measured almost 50 years ago. These and other known and well studied types of radiation are from electron related E/M sources. Light as we know it and of course CMBR, are all from electron activity. I have called the photons from electron activity E-Photons or EP for short. Light pressure from EP is perhaps measurable from the sun and other sources but it is very low in magnitude. For example, the light energy of all frequencies (EP only) received on earth (or near earth) from the sun per unit of area and per unit of time has been measured. This is typically of order 1500 watts per sq. m ($1.5KW/m^2$). This translates to total energy reaching the earth from the sun of order 1.5 K Joules /m^2/s. This is based on data from studies of the suns energy and needs further review but is adequate at this stage for first order analysis. The typical energy per photon (for a typical wavelength of solar EP radiation) is of order 10^{-18} J. Therefore an estimate of photon (EP) rate/density striking the earth from the sun would be of order 10^{21} photons per sq. m per second. Again translating this to momentum and pressure produces a figure that is far too low for gravity and hence for TOLG by orders of magnitude.

It seems as though there is a basic problem for TOLG with light pressure from EP. From a first order analysis (for a more detailed analysis see the chapter on light) it seems that EP is too low in energy and hence momentum to make light pressure gravity a reality. There are other problems with EP based gravity such as the solar effect. This would cause the earth to be blown away from the sun. And then there is the question as to how light impacts on nuclear matter (atoms, electrons, protons) and how it is deflected to impart momentum. These issues are also discussed more in the chapter on light. These and other factors lead me to the conclusion that EP in all its forms and from all sources is not sufficient for light pressure gravity. Hence for TOLG to be a reality I needed another source. The effects of the suns radiation pressure could be important in some other way and this may need to be taken into account but it is not part of TOLG. Perhaps there are some minor effects from solar wind and EP pressure but these are all considered to be of very low order with little or no impact on gravity through TOLG.

So a jump in thinking in physics was needed, the type of jump that creates new science. It seems that a new source of light or E/M radiation is necessary for CGBR and TOLG to be a reality. A type of radiation that has much higher energy (bigger rest mass!) and higher momentum, as well as other properties, than EP radiation, was required. Radiation that has not yet been observed, discovered, identified, detected, studied and controlled or even proposed, in the same sense that EP radiation has been. Wow, what a big ask. A quick review of the current model of all atomic and subatomic particles from hadrons and quarks with parameters such as spin, charm and strangeness shows no obvious candidates. Other Leptons, for example the Tau lepton, which is supposedly over three orders of magnitude heavier than the electron, but is not considered a form of E/M radiation or the neutrinos which supposedly have no mass and may be a form of radiation, may be possible sources. I am not familiar enough with the theory of the current subatomic particle model and more analysis is indeed required before these can be ruled out. However, I had serious doubts about these options as a source of high momentum radiation so I decided not to use these sources and to look elsewhere. There are many strange particles in the current model and I don't just mean quarks. There are many types of boson that have similar properties to light photons. Perhaps the closest idea that the current model seems to come to in regard to my proton light proposal, is the Higgs boson. It is an interesting part of the current proposed QCD atomic model of bosons, leptons and hadrons and was created to explain some unusual observations from atomic collision experiments and theoretical calculations. While it is supposedly a force carrier, whatever that means, it seems an ill-defined "particle" or boson of high energy with an undefined mass and speed. I am not an expert on the Higgs boson by a long shot (yes I say it again) but it seems an unlikely candidate for TOLG unless by chance it is the same as the PP in my theory. I will leave that for others to investigate.

There are two well known conventional electrical "particles" in the atom, the electron and the proton. The electron produces light (EP) so perhaps it is not unrealistic to assume that in a similar way, protons may produce light. We know that radiation/light (EP or electron photons) is related to the negatively charged atomic "particle" called the electron. So I proposed that the other positively charged atomic "particle", the proton, can also create E/M radiation or light (PP or proton photons). I have placed inverted commas around the term particle because I believe that there is still a lot more research and theoretical analysis is required before a more thorough explanation of the relationship between photons and their so called particle sources can be provided. I have previously called the electron photon an EP for short and now propose calling the proton photon a PP for short. The proposed properties of PP are addressed in more detail in the chapter on light. As the source of the PP, the proton is more massive or "heavier" so to speak than the electron (over three orders of magnitude), then it is not unreasonable to assume that a PP is also "heavier" in terms of rest mass and hence has much higher energy than an EP. Hence PP, which is still E/M radiation, is "heavy light" (more energy, more rest mass and more momentum). Therefore I almost called TOLG the Theory of Heavy Light Gravity (TOHLG) but perhaps this was just too cryptic for a title.

Just as EP come in a range of colours so to speak (the known E/M spectrum) and hence energies, PP may also come in more than one "colour" or energy level. Of course it is assumed that they would all be of much higher energy (orders of magnitude) that even the highest energy EP. The highest energy EP are believed to be of order 10^{-15} m in wavelength or 10^{24} Hz which can be translated to an energy level of order 10^{-10} J. Of course this is an area of considerable potential speculation until more research is carried out. It is possible that PP are more constrained than EP and have a limited but still quantised set of energy levels. New fields of proton theory and associated PP theory and technology as well as an associated area of quantum mechanics may be possible. At present little if anything is known about PP in a physical science sense, apart from the gravity effects of PP. However, it may soon be possible to detect some other forms of PP activity from high energy proton reactions, such as those that are being undertaken in the CERN (Proton/Hadron) accelerator. This is an area for considerable further study. It raises many more questions such as what is the relationship between PP and other particles in the current nuclear model, especially the so called Higgs Boson. How are they the related if at all?

LIGHT PRESSURE

It is well known that high speed particles such as a light photons have momentum and that when a particle impacts on a body in an elastic collision it imparts momentum due to the conservation of momentum. This change in momentum from a large number of particles causes a pressure. Now it is proposed that CGBR is just light and each CGBR photon (PP) therefore has energy and momentum and other properties according to current physics theory (Plank/Einstein) (i.e. it has a theoretical rest mass but has no "conventional mass"?). This is still a much debated area of quantum and classical physics but I believe TOLG will also support the idea of photons not really having any conventional mass and therefore also not being effected by conventional (Newtonian) gravity or being effected in any other "gravitational" way (General Relativity?).

The "mass" of a photon is derived from Plank's and Einstein's equations and the momentum is then obtained from the derived mass and speed of light. The correctness of these equations, relations and assumptions in this situation for CGBR needs to be carefully considered but is assumed to be a valid starting point. This momentum can be translated into pressure by considering the effect of momentum conservation in a collision with a body (in an elastic collision there is no energy loss or transfer of energy from the photon to the body but there will be momentum transfer). That is, a form of pressure comes from loss or transfer of momentum from CGBR photons per unit of area per unit of time to a body. A simple dimensional analysis shows how such a momentum change is related to pressure. Where m is mass, l is length and t is time.

Momentum
$$p = m * v$$
$$= m * l / t$$
$$= m * l * t^{-1}$$

Pressure
$$\rho = f / A = f * l^{-2}$$
$$= m * a * l^{-2}$$
$$= m * l * t^{-2} * l^{-2}$$
$$= p * t^{-1} * l^{-2}$$

So, for a source of proton photons (PP), each with momentum ρ_{PP}, and arriving at a rate of r per unit of time per unit of area and assuming all PP are travelling in the same direction (just for simplicity at this stage) and all single direction momentum is transferred in the collision to the matter in the impacting body, again just for simplicity, then the rate of momentum change for the given unit area per unit of time,

$$= \rho_{PP} * r * a$$
$$= \text{Pressure}$$

This is the basic concept of light pressure gravity. Of course more detailed analysis is required to derive an accurate quantification of real momentum transfer and hence pressure from PP, taking into account the actual energy of PP, the angles involved (incidence and deflection), the type of impact and the degree of momentum capture (related to "size" of particles involved and the capture or impact process), the degree of penetration of PP through any body/material and other factors. These are considered more in the chapter on light and will be the subject of more detailed study and analysis in ongoing research. This analysis will also enable the value of G, the gravitational constant, to be derived from other physical constants and factors and not from empirical studies.

So TOLG produces the same results in a static situation as Newtonian gravity which is given by the following formula;

$$F = \frac{G * m_1 * m_2}{d^2}$$

So calculating the force of attraction between the earth and the sun from these conventional gravity calculations (which are still basically accurate enough for non relativistic situations), where;

$\underline{G = 6.6 * 10^{-11} \, SI \; units}$
$m_1 = 6 * 10^{24}$ Kgm (earth mass) $m_2 = 2 * 10^{30}$ Kgm (sun mass)
$d = 1.5 * 10^{11}$ m

This gives a figure of approximately $3.5*10^{22}$ Kg m / sec / sec

Translating this to a pressure on the earth using say a 3dB type of equivalent size from the earths radius of say $5 * 10^6$ m, which gives an area of $6*10^{13}$ m², results in an equivalent pressure of around $7.5 *10^3$ Pascals. This is the effective net shadow pressure or the difference between the net pressures on each side of the earth due to the suns mutual shadowing with earth. The suns pressure difference can similarly be derived but of course it must be the same because as Newton stated and TOLG also shows the force of gravity on each body due to the other is the same. Then assuming say a 50% momentum transfer for each photon in the direction required, and using the previously derived momentum for CGBR photons, results in a rate of approximately $7.5*10^{20}$ photons/sec/m ². This can then be used to derive an energy rate or density for CGBR of about 25 KJ per cubic m. Of course these are all rough estimates for the purposes of checking overall model consistency. More analysis is required to review this estimate of the type of pressure likely per unit of area from a possible source of CGBR. This is addressed more in the chapter on light.

An allowance also needs to be made for the amount or degree of momentum transfer per photon (using conservation of momentum for CGBR photons) for a given type (density and volume) of matter. Does CGBR impact on atoms or only nucleons such as protons or neutrons? Depending on the answer to this (any many other questions) there may be differences/reductions) in this momentum transfer ratio. The absolute light pressure from CGBR also can't be derived directly from this figure as the amount of CGBR deflection by a body needs to be taken into account and this is still an area for further study. It could range from a reasonable amount for say a body like the sun (low/moderate percentage?) to a very low figure many orders of magnitude lower. This implies that general background gravity (CGBR) light pressure (without shadows) may be very high. This is another area for urgent further analysis. However I feel very confident that for reasonable assumptions about the possible amount (density) of CGBR, the frequencies and hence energy/momentum, the reflection/deflection made and rate of momentum transfer per unit of matter per unit of time (this is a difficult area!!) and perhaps other aspects, the theory will be validated to the extent possible. This will eventually allow a figure for G, the constant of gravity, to be derived from or related to other basic physical factors and constants.

SHADOWING

Just as matter interferes with light as we know it (EP) and can create shadows or dark areas where light is blocked, then it is assumed that every body with mass (nuclear material) reacts with or deflects PP (CGBR) to a degree. This reaction/deflection may create a shadow. This shadow may be on the body itself for non iso-cosmic radiation such as from a directional source like sunlight on the moon, or it may be cast on any another body. This shadowing is explained in the diagrams below.

If a body absorbs light (loss of energy) instead of just reflecting it away (change in momentum) there will be a resulting energy gain by the body. This would then produce a temperature rise and this would lead eventually to an energy loss in some other form of radiation. However, this doesn't seem to happen to our moon and any number of other "cold" bodies (Newtons apple), which are immersed in CGBR. Therefore, it would seem that CGBR is mostly reflected in a type of perfectly elastic collision. Hence the term "COLD CGBR" may be used due to its non energy transferring interactions.

So, it is assumed that CGBR must be totally (or almost completely) reflected by any (conventional atomic) body or nuclear matter and not absorbed. This is because the energy gained from any other non elastic collision and absorption would raise the temperature of the body and/ or change the nuclear structure. These factors would create an unstable situation, unless there is a matching form of energy release mechanism such as nuclear decay (more about this later). Perhaps there is some absorption but it is assumed at this stage that this would be minimal. Another area of interest is that even minor PP absorption and energy capture (and subsequent release) may be related to some form of atomic decay mechanism. This is an interesting subject for further study.

The mechanism of this reflection, whether it be at broad atomic level (unlikely) or at nucleus level (protons/neutrons) (more likely) is unknown at this stage. It is however proposed that it is at nucleus level. The PP "size" is considered to be much smaller than the EP "size" and hence PP should have a far greater degree of penetration in atomic matter than EP, in general. It is also proposed that the impact of PP on the nucleus would in general be elastic and would impart only momentum to the nucleus and hence the body in general. This "size" concept is addressed in more detail in the chapter on light. This almost total elastic collision process would prevent a body from creating a permanent solid shadow by itself from CGBR. This concept was explored at first as a possible shadow creating mechanism but was dismissed for this reason. Perhaps such an auto shadowing concept is similar in a way to Einstein's GR field. However, this non self shadowing property also makes all bodies seem to be invisible. For example, if a CGBR detector was used to search for (shadows caused by) an isolated body or mass, no shadow or variation in CGBR would be found anywhere near the body as there would always seem to be an equivalent amount of CGBR (PP) being reflected from one side and in the same direction as that which is blocked (reflected back) from the far side. This can be seen in a simplified way with the diagrams shown below.

An observer on the left side looking toward X would normally see the light beam (stream of photons) "a" as a simple incoming light source (CGBR) from the direction of X (shown as "a'" dotted). However with a (point) mass "m" blocking this light stream (the "a" photons) and reflecting it back in some direction (say as light beam "-a"), the observer will not see this light source. However, the observer can now see the light source "-b" as the reflected light beam from "b" that was not observed (could not have been observed as it was travelling away) before the point mass "m" was there. This new reflected light "beam" ("-b") now has the same path and appearance as one that would have been seen if the point body was not there (the dashed "a" beam). Therefore the point mass appears invisible for perfectly reflected ICR (CGBR) light.

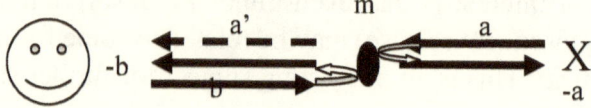

This is an idealised situation (approximation) but when based on an infinitely large number of CGBR PP's impacting on a body over a long period of time the result should be the same. In general, there would be an enormous number of photons going in virtually every direction possible and then being reflected in virtually every direction possible (like chaos!) but the overall result would be conceptually the same as shown by the simple analysis of one light beam. This analysis needs to be made more rigorous and extended to lager sizes and large particle counts, but the concept is sound and should be simply and linearly extendable. The result is that any body in isolation in a perfectly reflecting situation with CGBR would appear "invisible". That is because there is always reflected light coming from the same direction and in the same strength as that light which would have come if it wasn't there (i.e. from behind it), but was blocked and would have caused a shadow. Hence a body can not cast a shadow by itself. That is, it would not directly cast a solid shadow but would have the potential to cast a shadow on another body due to mutual shadowing. Therefore the only way a body can be seen (detected) or can create a shadow in perfectly reflected light (CGBR?) is when it is effected by another body (mutual effects).

THE MUTUAL SHADOW

So, if the real cause of gravity is CGBR pressure differences due to shadowing, what causes the shadow? From the above analysis, it seems unlikely to be due to a solid shadow caused by an isolated body due to light absorption. However when two bodies are considered together, mutual shadowing is created. That is, there is a form of shadow created on each body by the other when two bodies are "near" each other, or any distance apart for that matter. Any two bodies create a mutual shadow between themselves due to the blocking of light (CGBR) between them. Using the point mass example again, this time with two point masses A and B as shown below, the mutual shadowing concept is explained.

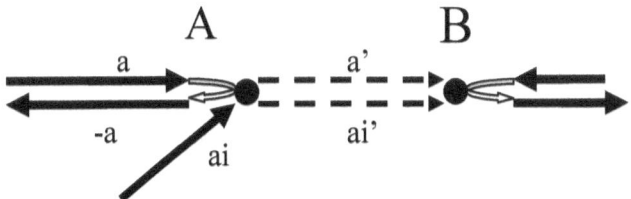

Beam "a" would have impacted on body "B" but with body "A" present, beam "a" is reflected back as beam "-a" by body "A". Therefore beam "a" beyond body "A" (shown as dotted "a'") is blocked from travelling to and causing "pressure on body "B". This creates a form of shadow on body "B" from body "A" due to light stream "a". There is no equivalent reflected beam to replace beam "a" as was used in the above non self shadowing description, as body "B" blocks light from travelling to "A" and being reflected back to "B" to create the replacement beam.

This situation is also matched by a beam coming from body "B" that would normally fall on body "A" but is now blocked by body "B". Therefore body "B" creates a form of shadow on body "A". While again this is a simplification of the total picture, it provides a general indication of the mutual shadowing effect between any two bodies. Of course the real situation is a dynamic moving one with all bodies moving in various/ different directions with varying speeds. Also, all PP directions are possible. Therefore a more extensive mathematical analysis is required to derive the total accumulated effect of all directions and reflections but the effects should be similar to the simple single point analysis above. A more detailed analysis is beyond the scope of this initial proposal.

While there would be others photons travelling in the direction of beam a, from other directions and other deflections from body A such as beam ai creating beam ai', and these would provide some pressure on body B in the direction from body A, the overall result would be a reduction in net light pressure in the direction of the shadowing body. So if there is no source of new light (CGBR) between the two bodies, this mutual shadowing results in a net pressure (CGBR light pressure) difference on each body in the direction toward the other body which creates the shadow and vice versa. This causes a force on each body pushing it toward the other body to bring the two bodies together. This is what causes the force of gravity! TOLG!, so simple and brilliant!

Of course "near" in this sense is a general concept and means any distance. The effect of pressure due to shadowing is directly proportional to the amount of matter creating the mutual shadow (mass of each body) and the inverse square of the distance between them (as per solid angle analysis below). Also as a body moves, the shadow also "moves" according to the speed of light and this means that the speed of gravity is therefore the same as the speed of light. This is an important point that Newton missed but is, of course, part of General Relativity. The impact of the speed of gravity being the speed of light on various gravity situations is addressed in the chapter on impacts of TOLG.

As CGBR is general background radiation from distant cosmic sources (discussed later), it is assumed that there would be no major sources of CGBR from our planets and probably most bodies in our solar system. Therefore bodies which have a gravity attraction, such as within the solar system and hence between the earth and the moon and also between the earth and the sun, create mutual CGBR shadows of the required strength. For gravity between the sun and planets it is generally assumed that the sun is not a source (large source?) of CGBR (see the chapter on light). If it is, then the mass of the sun may have been underestimated, more about this later.

If TOLG is based on CGBR collision activity at sub atomic level, which it most surely must be, then shadowing must also be at sub atomic or even sub nucleus level. The process must be a factor in deriving the shadow strength or density so to speak between any two bodies, taking into account the change in strength or direction of CGBR due to any other nearby bodies creating mutual shadows with the two bodies in question. This cumulative subatomic effect needs to be taken into account in TOLG. Another critical factor in this analysis is the degree of penetration of CGBR through bodies such as our earth and even the sun. It is likely the CGBR has a very high level of penetration even through such bodies as our earth. This is addressed in more detail in the chapter on light.

What happens when bodies (atoms) are in line? Does the shadowing compound or does the closest body (depending on size) prevent other more distant bodies from casting a shadow as well. This effect is called shielding and has been a factor previously used to criticise gravity theories based on pressure. Is shielding in fact, simply a restatement of the question of a body having different dimensions in the direction of the other body which is creating the mutual shadow? It may also be related to the type of CGBR momentum transfer or capture process that is a basic component of TOLG.

SHIELDING

I have included shielding in this section as it is a critical issue in shadowing gravity. It is addressed more in the section on problems with TOLG but here are some brief comments. The problem is caused by the potential reduction in the amount of shadowing produced by two pieces of aligned matter. The net shadowing may seem to be less than the sum of the shadowing that each of them would produce separately, due to overlap of shadows. Therefore the addition of matter may not result in a direct proportional increase in the strength of shadowing or in effective gravitational mass. This effect is called gravitational shielding. To solve this problem it has been suggested that the shielding effect is so small as to be unmeasurable. This of course requires that the cross-section of matter must be extremely small in realation to the collision process. This places a very high limit on the intensity of the radiation (photon energy/momentum) required to produce the observed force of gravity. Perhaps these factors are not unrealistic in TOLG.

The problem of shielding may be better understood by the concept of having more than two point particles (particles that are impacted by

CGBR) in a line at the time a photon is deflected by one of the "outer" ones. The "middle" particle therefore shields the other from the shadowing effect. A possible reason why shielding or interference in general may not be considered a real problem is that the probability of such alignment at the time a photon impacts the first particle and would have impacted the others, is so low that it can be ignored. However, as the overall effect is a large scale summation of such activities, the shielding error may also accumulate to a significant amount. This new proposal (TOLG) is based on light interaction (pressure) at neucleus level (protons/neutrons) from extremely short wavelength radiation (PP). These factors could be said to remove the problem of shielding apart from a small error. The real effect of shielding may be to create a small difference between gravitational mass (photon impact pressure) and real mass (sum of particles). This area has been the subject of much theory and research (inconclusive? unmeasurable?) but any form of gravitational shielding violates the equivalence principal which states the two types of mass are identical and this is inconsistent with G. R. Perhaps this issue will be better understood through this new theory. However, the supposed violation of the G.R. equivalence principal still needs to be addressed, if in fact it is appropriate.

THE INVERSE SQUARE LAW

The reason the inverse square law applies to gravity and hence TOLG, is related to geometry and shadowing. The basis of the inverse square law is simply the application of the concept of a solid angle. The concept of a solid angle was developed to handle things like flux of fields or rays of light emanating from a point source. If they rays are not isotropic but are restricted or focused in some way then they will not be evenly spread out over an enveloping shape wrapped around the source. For those not familiar with this concept perhaps a good example of this is a light house. The light points out in a narrow beam that is almost horizontal and rotates slowly around. If a capture device is moved around in space it will see different amounts of light ranging from almost none directly above or below the lighthouse to a larger amount in a horizontal direction away from the light house, which periodically flashes. For a uniform or isotropic point source, the amount of captured light for a given capture area would be the same in any position around the source. Now the solid angle of the captured light is related to the area of the capture device. Or in reverse it is the amount of area that the capturing device projects onto a unit sphere centred on the reference point.

Imagine an object and a point in space near to but outside the object. The solid angle is related to the area of the shadow cast by the object when projected onto a point (the reverse of the light captured by the object from a source at the point in question). It is easy to relate this to a shadow concept which is a key concept in TOLG. The inverse square law applies to the area of the solid angle and so to the shadowing effect which directly relates to CGBR pressure and hence TOLG force. The solid angle and resulting inverse square law concept is shown in the figure below.

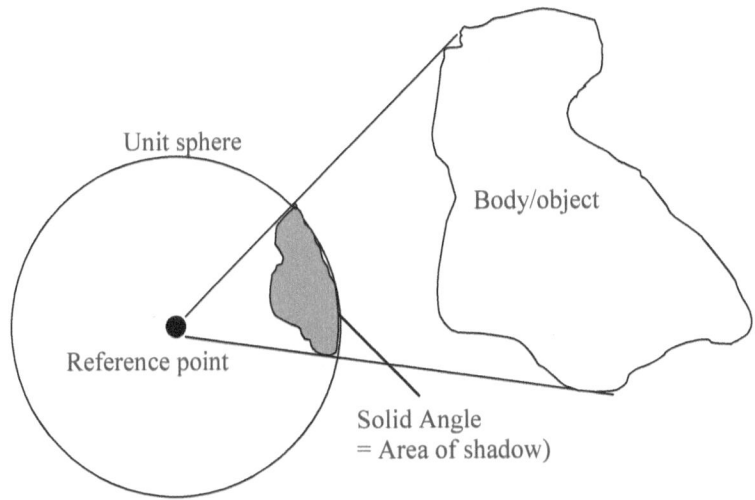

As the object is moved toward or away from the reference point the solid angle changes. It increases as the body moves toward the point, reaching a maximum of unity when the body completely encloses the reference point. It decreases as the body moves away, decreasing to zero at an infinite distance. Of course both these extremes have shadowing and hence TOLG gravity implications as well. The rate of change of the solid angle (area cast on a unit sphere) varies as the inverse square of the change of distance of the body from the reference point. The significance of this is that the inverse square law of gravity is simply explained in TOLG using the solid angle concept and mutual shadowing.

Of course, the concept should be linearly expandable and completely invariant to any shape, size, density, or any other factors that may be relevant to gravity analysis. So if any body is broken up into any number of component parts and the analysis is applied to each part, the same results apply. The TOLG shadowing and solid angle based gravity concept is simply a summation of all the gravity (shadowing) effects of all the component parts of any body on itself and any other body. It is a simple linear summation as expected. So shadowing or mutual shadowing follows the inverse square law of distance between the two bodies which is basic to gravity.

SUMMARY

This is a simplified analysis of the basic components of TOLG. It is not a complete and thorough theoretical and mathematical analysis of all aspects of the theory. Of course most physicists will say it is not rigorous enough for a scientific theory and yes a much more rigorous approach is of course required but I believe this is only a matter of time (and money?). However, the basic scientific ideas behind TOLG and the assumptions related to them are simple, sound and realistic. They are scientifically as consistent as many arguments that have been used before to justify many previous theories of everything. The main areas of any potential uncertainty and hence key areas for ongoing interest are CGBR and its sources and behaviour, the large scale summation of small collision effects and that of shielding. Everything else seems obvious to me.

The next step is to review all the assumptions and derivations and carry out more detailed analysis of this theory (TOLG). An overall theoretical model needs to be produced for computer analysis as I believe the TOLG model may be beyond the mathematics available today. This question alone will require extensive effort by many brilliant minds and perhaps the maths of General Relativity using tensors (which seems to produce the best gravity calculations in most cases) may still be appropriate. When the mathematics or the model is sound enough, a good confirmation task is to apply TOLG to some common gravity problems such as the unusual orbit (perihelion) of Mercury. TOLG is based on light (CGBR) so it therefore has a latency or delay due to the speed of light. This is what causes the unusual orbital behaviour of Mercury. This factor may also explain some other minor orbital differences in other solar system bodies that Newtonian gravity doesn't explain but G.R. does. But these cases alone don't prove TOLG.

The big question is how to fully prove TOLG is the correct gravity theory. Perhaps this can only be done with the detection of PP light. The Hadron accelerator at CERN may have the capability to accelerate protons to high enough energies to create PP and hence may be able to help find them. Other areas for review are the implications of TOLG on cosmic research and the behaviour of large cosmic bodies and on issues such as the Big Bang theory and Black Holes. Another challenge between TOLG and G.R. is the bending of light. In TOLG, gravity is created by light and as far as we know light doesn't bend other light. So perhaps there is another explanation for this observed phenomenon and I believe I know what it is. This and related issues are also addressed in more detail in other chapters. All these aspects need more rigorous analysis but I am sure some interesting new outcomes will be produced from the study and application of TOLG. Please read on to other chapters for more information and ideas on how this may be carried out. Any other ideas or comments on TOLG would be greatly appreciated but please keep them simple and clean.

CHAPTER 5

Problems and Proof

Any new theory of science, especially one that is not consistent with the accepted or standard model needs to be rigorously considered and proved. This may not be easy, especially if it is found to have problems or unknowns associated with it. That is not to say that a standard model or standard theory is always perfect, far from it. It is usually just the one that best fits the data and has a reasonable scientific framework behind it. As is often the case with standard models, they may start simple and explain most existing observations but as science and technology catches up, more and more problems or inconsistencies arise and the models become more complex. As models or theories are "improved" this often leads to their downfall. They make predictions or set limitations that constrain them and may drive them into obsolescence. This is not necessarily a bad thing as long as a better model results. Unfortunately two things seem to happen. Firstly, existing models are often supported well beyond their use by date especially by those with a vested interest. Secondly a major rewind or retraction of some existing "scientific" theory is often required before a new model can be established. Does this mean that even TOLG may eventually be superseded? Perhaps I had better stop and wait for a while!

This new theory of gravity has to prove itself to be worthy of replacing existing theories. It has to stand up to scientific scrutiny and hopefully solve more problems than it creates. Or at least it must be able to provide as many if not more answers than the standard model but with a simpler approach. I believe this new theory of gravity does just that. It is based on a simple concept and solves or will be shown to solve all known gravity problems just as well as the standard model does now. This new theory of gravity is well thought out and I am sure will eventually be proven to be scientifically correct. But at this stage it is in a state of development and yes, a number of questions still need to be answered before it can even be considered, let alone be accepted by the scientific community. However, this new theory of gravity faces greater hurdles than new theories in many other areas of science. Perhaps the most serious challenge is; why so late? Why haven't other famous physicists thought of it before? Why is this new theory being proposed by me only now after long support and apparent success of existing theories? I hope to answer this and the many other questions that arise and set the scene for serious consideration and then full acceptance of my new theory.

Perhaps the best way to start is to discuss the standard model of gravity. Does the existing theory or standard model work and provide correct answers. The answer to this of course is generally—yes. That is why it is the standard model. But ask the question a different way such as is the current model simple and understandable and is it always correct? The answer to this must be no. The current standard model is a mix of Newtonian gravity for simple situations with G.R picking up the hard stuff. Newton's law works well for most earth bound problems and then also provides a good result for many others. However for some well known situations Newtons law doesn't always work such as for calculating the correct perihelion advance of Mercury. The extended theory using G.R. supposedly solves this problem due primarily to the light speed of gravity and it works well for most other situations even if the mathematics is difficult. But there are situations that even G.R. can't handle and if so why? And G.R. does need a fudge factor called the cosmological constant. These types of problems are reviewed as they will most likely be significant to the acceptance of a new gravity theory. In fact it is important that a new theory can solve outstanding issues with existing theories.

Serious questions have previously been raised about shadow based gravity theories of which TOLG is one. Some of these were discussed and hopefully answered in the previous chapters but are raised again for completeness. While some attempt is made to solve these problems, the main aim at this stage is to put them into perspective. I wanted to find out if there are any fundamental and perhaps unsolvable problems or ones that require a solution which may severely damage the new theory? At this early stage of development I don't think there are any such issues. But in any case I am sure they will be solved following the development of a comprehensive mathematical model for TOLG. I will continue development until TOLG is "sounder" than all other gravity theories and fixes all the unknowns and is good enough to live with.

So the standard model may not be broken but it is not working perfectly. The errors may only be minor ones in the overall scheme of things but nonetheless some have been identified. Perhaps one of the fundamental problems with the standard model is that the cause of gravity is still a mystery. And the gravity constant G (both Newton's heuristic derivation and Einstein's simple copying) is not derived from any other more fundamental physical constants. It was merely created to do the job in the gravity equations. Is it a special property of matter and space/time or will it eventually be related to other more fundamental constants in an extension of G.R? In due course TOLG will enable this parameter to be derived from other more fundamental constants associated with atomic theory and light theory. But before addressing the problems of the current model or of TOLG in more detail, let's examine some key scientific questions for a new theory. These are not tests in the proof sense but are the types of questions that are usually asked of any new scientific theory.

1. **Is the problem worth solving?**
   The answer to this must surely be yes. The "problem" of gravity has already been worked on by the greatest physicists that ever lived so it must have been worth solving. Now I believe it can still be improved but I am not including myself in the category of great physicist yet. My new theory will add support to the existing foundations of gravity but it will also provide a more fundamental description for it and give a simple explanation for its cause. It should also answer outstanding gravity questions in due course.

2. **Does the current standard theory work?**
   While the answer to this question is a qualified yes there are still some problems in special situations. And there is a fundamental issue regarding at least one component of existing theory and that is; what

is the real cause of gravity? Existing theories both Newtonian and G.R. don't seem to have a satisfactory explanation for it, at least as far as I am concerned, and possibly also as far as many others are concerned, including some physicists.

3. **Is the new theory scientifically and technically sound?**
   The answer to this in my view is of course yes. However, this needs to be agreed to by the scientific community. Judgement or assessment can be based on this initial work but it may need to await detection of PP light and a comprehensive mathematical model of the new theory before a final decision can be made.

4. **Is the new theory based on new science?**
   The answer to this is both no and yes. The concept of pressure gravity and shadowing is not new. It has been around since Newton's time. But the cause of the pressure is new, being a new source or type of light. Gravity light is a completely new idea and perhaps it will be my main claim to fame.

5. **Is the new theory provable?**
   At this stage the new theory relies on an idea for a new type of light that has not been scientifically proven or identified. This is one of the main weaknesses of the theory and is the subject of intensive ongoing research by me and very soon I am sure by many others. It will be "found" soon, I am certain of that and when it is, my new theory will also be practically proven.

6. **Does the new theory lead to any new discoveries?**
   Yes it does, or perhaps I should say yes it should, a new type of light and then perhaps a completely new field of physics. It will also have major implications on many areas of gravity physics especially at cosmic and also at atomic levels because it challenges many aspects of other theories in these areas.

7. **What if any direct benefits will arise from accepting this new theory?**
   As mentioned before, unfortunately I can't come up with any ways of making money from my idea just yet but I am still working on it. I have already suggested some potential areas of application such as

a proton microscope, proton light lasers just like electron light lasers and perhaps one day an antigravity application. But these are at best a long way off and at worst beyond any technology. There may be some indirect spin offs in areas such as helping solve the problems of fusion research for cheaper power but I am only guessing and hoping here.

There may be others questions and answers but these will do for now to prove it is worth continuing research into TOLG. So now let's look at some specific problems associated with TOLG. Yes there may be others but this is a start.

GRAVITY LIGHT (CGBR)

The major challenge to TOLG is associated with the light or CGBR which creates a pressure difference that produces a gravity force. Where does this light come from, how is it produced and how does it behave? How does CGBR cause pressure and how does it impact on atoms? What are the effects of impact or collision at atomic level and is it simply reflected? Are these collisions elastic or inelastic? Then there are the problems associated with shadowing including the issue of shielding. And how far does PP or CGBR light penetrate matter, especially very large bodies such as the sun? There are also problems associated with the fast but finite light speed of gravity. It causes latency or delay in action and hence aberration.

Background light of various types and forms is not a new concept. Visible light from the cosmos has been observed since time began so to speak. Olbers paradox suggests that the night sky should show full starlight but it doesn't because of the density of matter in space, both light emitting and light absorbing. And then there is aging red shift which reduces the wavelength of all light to eventually become CMBR but that is another issue. A different type of background radiation, Cosmic Microwave Background Radiation (CMBR), has been studied at length since it was first detected and measured almost 50 years ago. These types of background radiation or light as we know it (Jim) and of course CMBR, are all from electron activity. This activity creates or consumes electron photons. I have called the photons from electron activity E-Photons or EP for short. The behaviour of EP is well understood. Light pressure from

EP from the sun can be calculated but it is very low in magnitude, too low for gravity. That's when I discovered PP or gravity light.

LIGHT PRESSURE

For example, the light energy of all EP frequencies received on earth from the sun per unit of area and per unit of time has been measured. This is typically of order 1000 watts per sq. m (1 KW/m^2). This translates to total energy reaching the earth from the sun of order 1K Joules $/m^2/s$. This is based on data from studies of the suns energy and needs further review but is adequate at this stage for first order analysis. The typical energy per photon (for a typical wavelength of solar EP radiation) is of order 10^{-18} J. Therefore an estimate of photon (EP) rate/density striking the earth from the sun would be of order 10^{21} photons per sq. m per second. From a first order analysis (for a more detailed analysis see the chapter on light) it seems that EP is too low in energy and hence momentum to make light pressure gravity a reality. So there is a basic problem for TOLG with EP light and EP pressure. There are other problems with EP based gravity such as the solar effect; that is the earth would be blown away from the sun if CGBR was EP based.

That is where my new light comes in. It is based on a new type of physics, proton based light or proton photons. I have called electron photons EP for short and call the proton photons PP for short. The proposed properties of PP are addressed in more detail in the chapter on light. As the proton is more massive or "heavier" so to speak than the electron (over three orders of magnitude), then it is not unreasonable to assume that a PP is also "heavier" in terms of rest mass and hence has much higher energy than an EP. Hence PP, which is still E/M radiation, is "heavy light" (more energy, more rest mass, more momentum). Therefore I almost called it the Theory of Heavy Light Gravity (TOHLG) but perhaps this was just too cryptic for a title. The science and technology of EP as well as its effects on our lives, is very well known and understood. Our lives are driven by the suns EP radiation (light, etc.). We also control and use EP technology for so many aspects of daily living. However, it seems we have no direct experience with the science or technology of PP or its impacts on us. But as TOLG shows, PP is very close to us in a direct way because PP causes gravity. Gravity is very significant to our lives and so we are as dependant on PP radiation as we are on EP radiation.

The real question is how to find out more about (discover) PP. This is an important question for research.

Just as EP comes in a range of colours so to speak (the known E/M spectrum) and hence energies, PP may also come in more than one "colour" or energy level although it is assumed that they would all be of much higher energy (orders of magnitude) that even the highest energy EP. The highest energy EP are believed to be of order 10^{-15} m in wavelength or 10^{24} Hz which can be translated to an energy level of order 10^{-10} J. From an order of magnitude analysis it is likely that PP have frequencies of order $\sim 10^{27}$ Hz or greater, with corresponding wavelengths of $\sim 10^{-18}$ m or shorter and hence energies of $\sim 10^{-7}$ J or approx. 50 GeV or more. It is possible that PP are more limited than EP and have a more restricted but still quantised set of energy levels. At present nothing is anything is known about PP in a physical sense, apart from their gravity effects. However, it may soon be possible to detect PP activity from high energy proton reactions, such as those that are being undertaken in accelerators. This is an area for study and raises questions such as what is the relationship between PP and other particles in the current nuclear model such as the so called Higgs Boson? Are they the same or are they related and if so how? The idea of gravity photons, so to speak, is not really new. It has previously been proposed that gravity is "caused" by a field or light like particles These "things" that cause gravity were called gravitons and still are I believe. But the view was (still is?) that gravitons somehow carry an attractive force (a kind of suction force) whereas CGBR "gravitons" simply cause a pressure due to momentum. But perhaps I shouldn't use the term graviton and should stick with my term of CGBR.

The absolute light pressure from CGBR can't simply be derived from an average rate figure as the angles of incidence and deflection need to be taken into account. Also the amount of CGBR momentum transfer due to deflection by a body needs to be taken into account. However I feel very confident that for reasonable assumptions about the possible rate or density of CGBR, the wavelengths and hence energy/momentum, the reflection/deflection mode and rate of momentum transfer per unit of matter per unit of time (this is a difficult area!!) and perhaps other aspects, the theory will be validated to the extent possible. This will eventually allow the constant of gravity, G, to be related to other basic physical factors and constants. But to get a better understanding of the effects of the light deflection process and how it causes a gravity pressure difference and also how shielding arises, lets look at the collision, penetration and shadowing concepts in more detail and try to identify any problems.

COLLISION AND PENETRATION

One area for consideration is the process of CGBR photon (PP) collision at atomic level. Does CGBR impact on complete atoms or only on nucleons or hadrons such as protons or neutrons? Depending on the answer to this (any many other questions) there may be differences/reductions) in the momentum transfer ratio. And if there is an element of capture, even if it is only for a short period of time until the energy is re-emitted, what does this mean for the temperature. Does it cause the body to heat up? Is this a possible cause for the apparent heating up of the earth beyond what is normally expected from current solar based earth heating and cooling models? And of course do proton energy levels effect temperature in the same way that electron energy levels seem to? If there is a degree of capture it doesn't seem to cause heating as shown by our cold moon and any number of other "cold" bodies (Newtons apple), which are immersed in CGBR. Perhaps there is some absorption but it is assumed at this stage that this would be minimal. This PP absorption and energy capture (and subsequent release) may be related to some form of atomic activity or decay mechanism which releases the energy before the body heats up. Hence the term "COLD CGBR" could be used for PP due to its non energy transferring collisions and interactions. So, it is assumed that CGBR is mostly reflected in a type of perfectly elastic collision.

The mechanism of this collision and reflection is proposed as being at nucleus or hadron level. The simple reason for this is the "size" of PP. They are considered to be much smaller than the EP "size". One outcome of this size is that PP should have a far greater degree of penetration in atomic matter than EP, in general. This "size" concept is addressed in more detail in the chapter on light. It is also proposed that the impact of PP on the nucleus would be elastic and would impart momentum to the nucleus and hence the atom and to the body in general. Also how far can CGBR penetrate a body or what is the mean path length before a collision? Is it due to PP "size" and the relative size of nuclei compared to atomic sizes and atomic density in matter? And hence what is the probability of CGBR passing through a body? It could range from a low amount or zero for say a large massive body like the sun to a very large figure many orders of magnitude higher for a small body such as a 1Kgm mass. This implies that the rate of general background PP light (CGBR) may be very high.

MUTUAL SHADOWING

If the CGBR light is not iso-cosmic but has greater strength from one direction then the body will create a shadow on itself away from that direction. This is like the shadow we see on the moon from directional sun light. But if the CGBR is fully iso-cosmic and equal from all directions then a body can't cast a shadow on itself. But it can cast a mutual shadow on another body. Perhaps such a potential shadowing concept is similar in a way to Einstein's G.R. field concept. Therefore an almost total elastic collision process together with iso-cosmic CGBR would prevent a body from creating a permanent shadow by itself. This non self shadowing property also makes all bodies seem invisible in CGBR light. For example, if a CGBR detector was used to search for shadows caused by an isolated body or mass, no shadow or variation in CGBR would be found anywhere near the body. There would always seem to be an equivalent amount of CGBR (PP) in an place near the body (but not inside it?). Light would be reflected from one side and in the same direction as light from the far side which is blocked by the body.

But when two bodies are considered together, mutual shadowing is created. That is, there is a form of shadow created on each body by the other when two bodies are "near" each other, or any distance apart for that matter. Any two bodies create a mutual shadow between themselves due to the blocking of light (CGBR) between them.

While there would be others photons travelling in the direction of beam "a", from other directions and other deflections from body "A" such as beam "ai" creating beam "ai'", and these would provide some pressure on body "B" in the direction from body "A", the overall result would be a reduction in net light pressure in the direction of the shadowing body. So if there is no source of new light between the two bodies, this mutual shadowing results in a net pressure difference on each body in the direction toward the other body which creates the shadow and vice versa. This causes a force on each body pushing it toward the other body to bring the two bodies together. This is what causes gravity!

Another critical factor in this analysis is the degree of penetration of CGBR through bodies such as our earth and even the sun. It is likely the CGBR has a very high level of penetration even through such bodies as our earth. What happens when bodies (atoms) are in line? Does the shadowing compound or does the closest body (depending on size) prevent other more distant bodies from casting a shadow as well. This effect is called shielding and has been previously used to criticise gravity theories based on pressure.

SHIELDING

In a sense shielding means that the combined effects of multiple bodies may not be same as the simple cumulative effect of each body. The problem of shielding in TOLG may be better understood by the concept of having more than two point particles (particles that are impacted by CGBR) in a line at the same time a photon is deflected by the "outer" one. The "middle" particle therefore shields the other from the shadowing effect. A possible reason why shielding or interference in general may not be considered a real problem is that the probability of such alignment at the time a photon impacts the first particle and would have impacted the others, is so low that it can be ignored. However, as the overall effect is a large scale summation of such activities, the shielding error may also accumulate.

Shielding is explained in the diagram below which uses three bodies (particles) for simplicity. Of course shielding is a multi particle (body) problem but an examination of the three particle situation will provide a useful explanation of the issue.

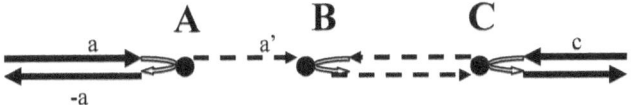

Assume all three particles "A", "B" and "C" are in line when PP "a" strikes "A" and also when PP "c" strikes "C". As all particles are assumed to be non stationary, which is a reasonable assumption for any real atomic particles, especially those unrelated, then the alignment assumption is a big ask! (i.e. has very low probability). The problem is really a dynamic one with all bodies moving in varying directions and PP possible from any directions and generally in equal strengths. If the alignment assumption is in any way reasonable then the following analysis applies. PP beam "a" would have impacted on body "B" but with body "A" present, beam "a" is reflected back as beam "-a". Therefore beam "a" is blocked from travelling beyond body "A" (shown as dotted "a'") and from causing pressure on "B" and also on "C". So there is no equivalent "a'" beam to create pressure on "B" or "C". So "A" creates a shadow on "B" and "C". Similarly, "C" and beam "c" create a shadow on "A" and "B". But in effect "A" shields "B" and "C" shields "B". Therefore it appears as though the middle body "B" is not able to create a shadow on "A" from beam "c" or on "C" from beam "a". That is, body "B" is trapped in the middle so to speak and has no net shadowing effect on "A" or "C" and hence has no net pressure effect. It is left floating between the two. Of course this is exactly what would be expected for a body caught between the gravity of two other (equal) bodies such as "A" and "C". The only real effect is a gravity force (pressure difference due to shadowing) that forces "A" and "C" together. There is no gravity pressure on "B" toward A or C from light sources c or a.

If bodies "A" and "C" are not identical in terms of shadow power (i.e. different masses or size*density products), then there would be some net pressure on "B" toward the larger shadowing body. The point at which "B" is balanced is when it is at the shadowing centre (centre of mass/gravity) of two (or more) bodies. If it is anywhere on the CG plane between the two bodies then the only pressure (gravity) it would experience due to them would be to push it toward the central CG. In reality, "B" would cast a shadow (due to penetrating/reflecting CGBR) on both "A" and "C" and would therefore create a pressure force (gravity) force on these two bodies. This is based on the assumption that every body has some form of CGBR capture area and is not a true point source without dimensions.

The probability of any in line effect is perhaps very small and perhaps could be ignored especially if the bodies are any real distance apart (considerably more than typical atomic sizes). Perhaps this shows that shielding is a non issue. That is, shielding just creates more complex shadowing arrangements and doesn't change the overall shadowing effect of one body on another, regardless of whether other bodies are involved in any location. However this needs more detailed analysis. In general it should be possible to show that with any number of bodies in any position going in any direction, there will be mutual shadowing between any two of the bodies that is independent of any other body. Unless there is total alignment (for an extended period?) then shielding may not be significant. Perhaps this issue is related to the difference between the weak gravity force at long (more than atomic) distances and the strong nuclear forces which operates at shorter distances. However, another way of considering shielding is that it is just part of the shadowing process. In the model above, the shadow from A on C is also cast on B and from C on A is also cast on B. Particle A has no real information on the real source of its shadow and could assume it is from particle B. Also particle C has no real information on the source or location of its shadow causing body, apart from changes in strength from the inverse square law but would this apply in such linear situations. It can't tell if it is from A or B. It therefore seems that in this case shielding has no real impact on shadowing and hence on TOLG.

Another way of looking at shielding is that although matter is proposed as having low density, it could not be completely transparent, because if it were no collisions and hence pressure would occur and hence no gravitational force would exist. However, this non transparency also seems to cause a problem because it suggeste that the linear addition of matter does not result in a linear increase in the strength of shadowing or in gravitational force. To solve this problem it has been suggested that the shielding effect is so small on most bodies as to be unmeasurable. This of course requires that the active interaction density of matter must be extremely small. This places a very high value on the intensity of the radiation (photon energy/momentum) required to produce the observed force of gravity, and a very low limit on the "size" of light causing such force and a low limit on the size of the nucleus of an atom. Perhaps these factors are not unrealistic for CGBR and TOLG. This new proposal (TOLG) is based on interaction (pressure) at neuclus level (protons/ neutrons) from extremely short wavelength radiation (PP). These factors may solve the problem of shielding apart from a small error.

In another example, imagine two bodies of equal volume of matter or mass but with different shapes. From conventional gravity theory, they would both weigh the same or have the same gravitational force at the same distance from a gravitating body. This type of experiment could and has been carried out many times on earth and has been used to prove that body shape has no detectable impact on force of gravity to a high level of accuracy. But the shape of material in the direction of the gravity field would supposedly have an impact on gravity due to shielding. However, it seems as though, to a reasonable degree of accuracy, that both geometries produce the same gravity force (i.e. have the same weight). Hence, if shielding is a factor they must have the same amount of shielding. How can this be when obviously a long thick object should have much more shielding than a thin object of the same volume and hence the same matter? But this simple analysis ignores any angled CGBR effects and these need to be taken into account. This may show how multidirectional and dynamic radiation effects can overcome what seems to be a potential shielding difference between the two geometries.

Comparing two shapes with the same mass but different geometries is shown in the diagram below. The first is a thin flat plate with a large area, aligned laterally to the radial gravitational field lines. The second is a long thin rod aligned in parallel with the direction of the gravitating field ("radial gravitational field lines or CGBR"). For the thin plate, there would supposedly be less shielding than for the long rod. The thin plate would present less matter and hence provide less shadowing but it would be over a greater area. The thin rod would present more linear matter and hence produce a "darker" shadow but it would apply over a smaller area. The total volume and hence mass and the average distance should be the key factors in determining the total shadow intensity. The mass (for equal density matter) would be related to the volume of each body and this is derived from the product of the cross sectional area orthogonal to the gravity field and the depth of material in the direction of the field. The distance would be from the centre of gravity (CG).

SHIELDING COMPARISON

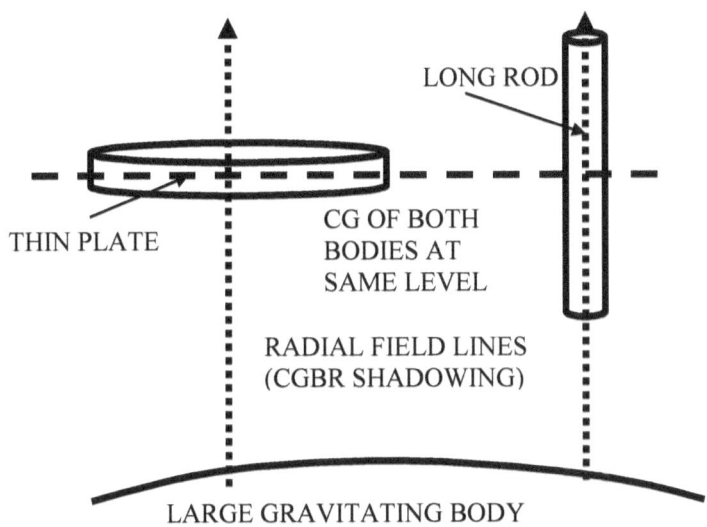

Would the two cases provide the same gravity effect? This is the shadowing and shielding question. Assuming both bodies have the same volume and hence mass do they produce the same shadowing and also the same shielding in TOLG? A possible effect of shielding may be to create an extremely small difference between gravitational mass (photon pressure) and real mass (atoms). This area has been the subject of much theory and research but the results to date seem to be inconclusive as there seems to have been no easy way of conducting measurements for large masses in an isolated or highly accurate way. According to existing physics theory, especially G.R. any form of gravitational shielding violates the equivalence principal and this is unacceptable. However is violation of the equivalence principal really such a fundamental issue.

MASS = WEIGHT - EQUIVALENCE

A basic tenet of G.R. is the equivalence of gravity weight and atomic mass. Many measurements have been made to test this theory. Most are based on a comparison of atomic mass (atomic count) and force of attraction between two large objects suspended very close together by long fine cables or on long rods. The atomic mass is estimated from volume and other atomic parameters. The gravity force between each body is then measured by extremely accurate and sensitive devices which measure the force of attraction between them. The two are then compared. However, all efforts of this type have not identified any difference between mass and gravity force (weight) for large but in reality relatively small masses. But what about differences in mass and weight for earth or sun sized bodies. The best ways of reviewing these large bodies have been based on estimates of structure, content and density. It is interesting to note that the estimate of the suns density to get the required weight and mass equivalence is quite low. TOLG may provide a new means of addressing these questions.

This raises the intersting question of why all bodies fall at the same rate. Well TOLG answers this easly. All bodies are made up of the same basic building blocks at atomic level. For TOLG the force of gravity is due to pressure from CGBR impacting on subatomic particles like protons and neutrons. Therefore as all atomic matter will suffer the same TOLG effects and hence any body made up of atoms will fall at the same rate. It is simply an integration of the effects of gravity from CGBR on many small particles.

DRAG

Drag was one of the main problems identified with early pressure gravity theories. It was considered a potential problem if the pressure causing fluid was something more physical than light. According to conventional shadow theories an isolated body is subjected to some form of drag if it is in motion relative to the unique isotropic frame of the pressure causing "fluid" (i.e., the frame in which the speed of the fluid or light is the same in all directions). This is due to the fact that, if a body is in motion, the particles striking the body from the front have a higher speed (relative to the body) than those striking the body from behind. This is simple Newtonian mechanics again. If it was true the effect would be to slow the bodies down and hence reduce the distance between the two bodies such as between the sun and the earth which would cause the earth to spiral into the sun and fortunately this hasn't happened yet. Early pressure theories were based on a fluid of ultramundane particles which naturally had conventional fluid like physical properties such as viscosity. When the idea was extended to more esoteric fluids such as a luminiferous aether or just ether, it was believed this would still create a drag if the ether had any real physical properties. Even after many experiments no such properties were detected so was it still considered a possibility?

For TOLG based on CGBR, there is no obvious way that CGBR light can create a drag on any body moving through it in any way. From Special Relativity, the speed of light including CGBR light is the same in all directions for any body, moving in any direction at any speed (below the speed of light). Therefore gravity based on light pressure could have no directional drag effects due to speed differences. But what about changes in the energy of PP in a directional sense due to movement through it? Would this have a pressure difference effect and hence a directional gravity drag effect? Didn't some physicists recently detect such an energy effect or a red/blue shift with CMBS? If movement really caused a change in the energy and hence momentum of the light in a directional way there may be some drag effects but this red/blue shift in CMBS as well as in CMBR is a challenge to me. It seems to contravene relativity but I need to give it more thought. I don't believe this is a problem with TOLG but will review it in further work.

SPEED OF GRAVITY

Some scientists suggested that gravity is faster than the speed of light! But in reality, the speed of gravity is the same as the speed of light. G.R. states it is and TOLG explains why this is so. But is the speed of gravity the same as the speed of conventional light or not? We know the speed of conventional light or EP light as I call it, from measurements and Maxwell's equations. It depends on properties of nature or perhaps I should say properties of the media. But it is also known that the speed of light through a medium depends to a certain extent on the frequency or wavelength of the light. The wavelength or frequency of light or of each photon is of course directly related to the energy of the photon. For a given media such as glass, longer wavelength light or lower energy photons are generally slower than shorter wavelength ones. Lower frequencies also bend more at interfaces with other media. The actual speed difference for different wavelengths of PP light is very minor but is sufficient to cause detectable interference and other side effects. This is the basis for diffraction and the rainbow effect that Newton studied so much.

Now PP light is much shorter in wavelength than EP light so does it travel faster? If so, then by how much and how does the speed of light change as a function of wavelength or photon energy. Is it due to changes in the parameters that determine EP light speed? Or more precisely how much do the parameters that determine light speed change in relation to wavelength? And just as importantly, why do these parameters of a media change as a function of wavelength? In a specific media, it may be possible to somehow relate this phenomenon to physical sizes of atomic material and the "capture" process for light. However, in a vacuum, what is the mechanism that causes this change? Perhaps there is no such thing as a pure vacuum without any matter (or energy) but then what is in a vacuum that causes the speed parameters to change as a function of photon energy? And if gravity is caused by mutual shadowing, what does the speed of gravity really mean? How can allowances be made for the speed of shadowing? Well these and related questions will be answered in the next exciting episode but I don't believe they are very significant. The speed difference between CGBR and conventional light is perhaps at best negligible and at worst very small.

ABERRATION (LATENCY)

One possible effect of a finite speed for gravity is orbital or gravitational aberration. If the gravity is moving at a finite speed, there is a time delay in the interactions between bodies (the transit time). If either or both of the bodies are moving this results in each body reacting to a previous position of the other, which creates another force component. It is proposed that this may act to push bodies away from each other. Many earlier pressure gravity theorists or others working on the speed of gravity such as Laplace, used this to prove that in order to maintain stable orbits, gravity must be much faster than the speed of light. This has also been suggested by many as a conclusive disproof of all pressure theories. In contrast, G.R. is consistent with the lack of appreciable aberration because even though it proposes that gravity propagates at the speed of light, the expected effect is supposedly cancelled by velocity-dependent terms. Of course there is no reason why this explanation can't apply to light pressure gravity as well. This needs to be addressed in more detail to solve this aberration or latency issue.

So if gravity acts at a finite speed there may be measurable effects on gravity in situations where either or both bodies are moving, especially if they are moving rapidly. Yes, I know this immediately raises the question of moving relative to what or relativity in general but this is an interesting observation which is addressed elsewhere. But let's continue based on using the background universe as a reasonable frame of reference just like Mach and Einstein? Aberration, or latency as I call it, is due the finite (light) speed of CGBR in TOLG. It is caused by the combined effects of a delay in the transmission of the gravity "signal", or the shadow in TOLG, from one body to another and the distance travelled by the bodies during this delay. It's a bit like shooting at a moving target, you need to shoot ahead of it. What would the effect of this be on TOLG. Well firstly it may make latency or aberration, a major issue. In fact we need to take account of the movements of all the bodies being considered.

In the simplest case let's look at our solar system with the sun and all the planets. Earth movement consists of several types. The first and most obvious is of course axial spin. This gives us our days and saves us from being destroyed by the sun, which would happen to one side of the earth if the earth didn't rotate or to all sides if it rotated slowly. This rotation results in an equatorial ground speed of about 1500 Km per hour (36000Km per 24 hours) or 400 m/sec. This high speed is unnoticeable by humans and is almost undetectable unless special devices are used such as sinks and plug holes. It also causes a reduction in weight at the equator compared to the poles but not by much. But the effect is enough to make the earth oblate or fatter in the middle just like me. But this rotation would have no impact on gravity between the earth and sun or the earth and any other body. It is purely a local experience and would not cause aberration or latency.

The next external movement that would be relevant to latency is the rotation of the earth around the sun. This annual rotation it results in a speed of about 100,000 Km per hour or 300 Km per second. Again this is not detectable by humans even though it is not a linear or inertial movement but a circular accelerating one. The next movement of the earth is one associated with our total solar system. Of course our solar system is not fixed in space and moves. The most obvious movement is due to the rotation of the Milky Way galaxy. We are not in the centre but nearer the edge so we are moving quite fast and the current estimate of this speed of the total solar system is about 220000 Km/sec. Hence the net maximum speed of a point on the earth is quite fantastic. Then there are of course other movements. Our galaxy is surely orbiting around other galaxy clusters or even larger cosmic masses such as super clusters of galaxies but I will not continue this line anymore. Suffice to say there is movement of the earth and the speeds are very high. As Galileo said about the earth; "and yet it moves" but if he only knew at what speed. But does this movement cause aberration, it seems not. Perhaps it is not large enough.

But there is possibly an even larger type of movement. According to the Big Bang theory, our complete Milky Way galaxy and therefore the sun are moving away from the "centre" of the Big Bang (I know there is supposedly no such thing as a centre of the big bang but let's continue for a while). Our sun must therefore be moving at some reasonable fraction of the speed of light relative to this "big bang centre" of the universe or at least relative to some other points of the current (past?) universe in the Big Bang model. For some reason these ideas of a centre of gravity and a reference point for the start of the Big Bang universe are not allowed but more about that later. The reason for this high speed movement is the Copernicus principle again. It simply says that our sun is just like any other sun or our galaxy is just like any other galaxy in general terms. And it has been "proven" from red shift observations that many suns (galaxies) are expanding outwards or should I say moving away from each other. So our sun must also be moving away from something at a speed which must be very fast indeed. This high speed movement of the sun and solar system alone would cause a significant gravity latency effect, and this should even show up in G.R. For example the sun could easily be moving at some reasonable fraction of the speed of light, say 10% of the speed of light, which is a totally realistic Big Bang based prediction. If so then it will have moved about $1.5*10^{10}$ metres or about 20 sun diameters in the time the suns light takes to reach the earth. Of course this is additional to all the other movements mentioned above and at some times these could be aligned. So if the Big Bang theory is correct the obvious interpretation that we are moving at a reasonable fraction of the speed of light follows, then there would be significant latency or aberration. But there isn't! This may need relativistic calculations to calculate the full effect but it would seem that it could be very high? So perhaps there is no such rapid expansion movement here or anywhere else! Another Big Bang challenge.

Another task is to apply TOLG to some common gravity problems such as the unusual orbit (perihelion) of Mercury. As TOLG is based on light (CGBR which therefore has an action latency or delay due to the speed of light, this may be the key factor differentiating TOLG from Newtons gravity which assumed gravity forces to be instant. This factor together with the speeds (orbital, other) of the bodies involved may explain the special orbit of Mercury as well as some other minor orbital differences in other solar system bodies that Newtonian gravity couldn't explain. Other areas for review are the larger cosmic issues such as the Big Bang theory and Black Holes. There are other challenges to current theory (G.R.?) such as the bending of light. In TOLG, gravity is created by light and as far as we know, light doesn't bend light so perhaps there is another explanation for this observed phenomenon. This and related issues are also addressed in more detail in the chapter on light. These aspects need more rigorous analysis but I am sure some interesting new outcomes will be produced from the study and application of TOLG.

SOME GRAVITY TESTS

Let's look at one particular gravity situation that is quite challenging, is that of gravity forces on an enclosed body. One interesting aspect of gravity is that a body that is inside another symmetrical hollow body or a spherical shell of material "feels" no gravity toward that shell. This is true no matter what position the internal body is in within the shell or how close it gets to any part of the shell. This comes out of Newtons laws using simple calculus. Although perhaps the effort he went through to develop his calculus (Fluxions) to derive this result was more complicated than the end result.

This can be interpreted in TOLG in terms of mutual shadowing. In TOLG this internal body situation implies that a shell can't cast a directional mutual shadow on an internal body and vice versa. Upon closer examination, however, the shell does cast a shadow on the internal body. What this means is that the level of CGBR is lower inside a body than outside it. So the internal shadow strength will be identical in all directions and should not change as the internal body within the shell changes position. The reasons for this are related to the concept of solid angle. A shell casts a fixed solid angle ($4\pi r^2$) on any internal body, regardless of geometry or location. So there would be iso-cosmic CGBR within any shell and hence there should be no directional CGBR and no pressure difference or force in any direction on the internal body due to the shell. So it follows that there would therefore be no net force to move the internal body in any direction, especially not toward the centre of the shell or toward the shell itself.

This outcome matches other gravity theories. Newton's gravity (and GR?) also proves that there is no net force on an internal body in such a situation, regardless of the position of the body. The internal body must also cast some type of mutual shadow from CGBR on the shell but again this would appear from similar considerations, to be equal in all directions. Hence it would not cause any net force on the external shell in any particular direction, no matter how close the internal body came to the shell.

Now if the shell and internal body are in an external gravity field or should I say CGBR shadow (from another nearby body) both the shell and the internal body should experience a gravity force toward the external body. The question is, how does TOLG cater for this situation? Of course it must produce the same results. But this immediately raises the shielding problem. The shell and the internal body must feel a pressure difference due to the mutual shadow of the external body and the external body must feel a pressure difference due to the mutual shadow of both the shell and the internal body. So if both the shell and the internal body deflect some CGBR that was in the direction of the external body then this shadow strength is simply due to the summation of the matter in both bodies. However, for the mutual shadow on the external body, whose strength is related to the mass (matter) of the external body, it must then have an effect on both the shell and the internal body. This situation is complicated because there are now three bodies casting shadows.

Perhaps this can best be explained in the following diagrams.

SITUATION 1 – SHELL WITH NO EXTERNAL BODY

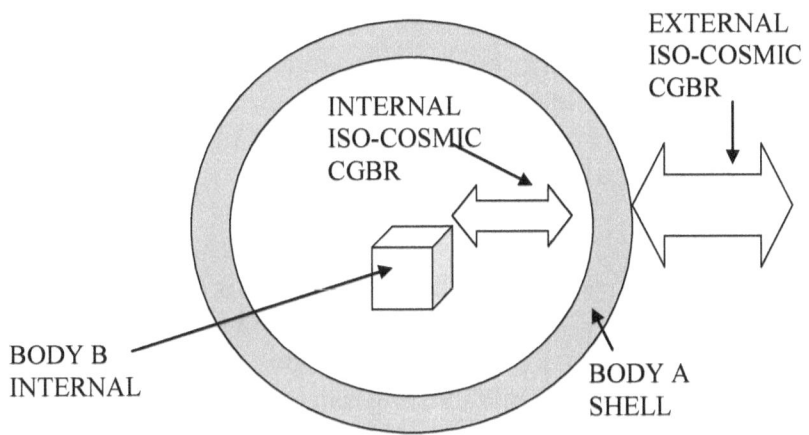

EXTERNAL
ISO-COSMIC
CGBR

INTERNAL
ISO-COSMIC
CGBR

BODY B
INTERNAL

BODY A
SHELL

SITUATION 2 –SHELL WITH EXTERNAL BODY

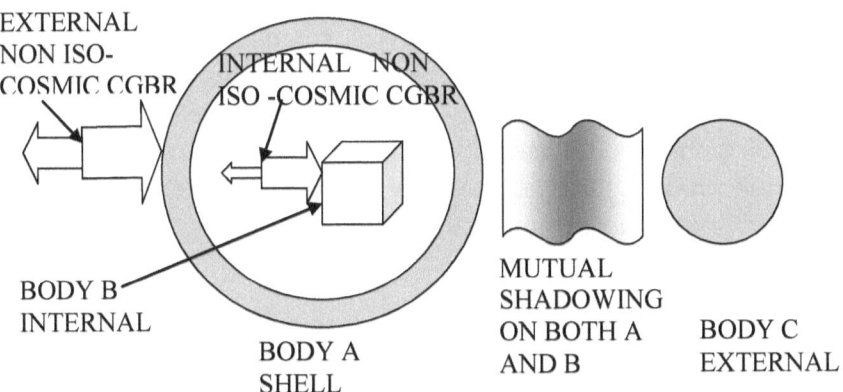

EXTERNAL
NON ISO-
COSMIC CGBR

INTERNAL   NON
ISO -COSMIC CGBR

BODY B
INTERNAL

BODY A
SHELL

MUTUAL
SHADOWING
ON BOTH A
AND B

BODY C
EXTERNAL

MATHEMATICAL MODEL

Perhaps the most fundamental problem with TOLG is that a full analytical solution and a comprehensive mathematical model have yet to be developed. While a simple model has been proposed as to how TOLG works and a first order analysis has been carried out to prove the concept in general, much more work is required in this area. The development of a full analytical model presents some real challenges not just to me with my limited maths ability but in general terms. Unfortunately I don't have anywhere near the maths abilities of Einstein or even Newton. I am working on this problem but may run out of time. Any assistance would be greatly appreciated I say jokingly.

The problem is associated with integrating the very minor effects of a very large number of many small photons travelling in every conceivable direction and creating dynamic shadows on moving objects. A form of summation on a large scale and over time of almost infinitely many small events will be required. While existing calculus methods may apply, calculus is used when no finite limits apply and this may not be the case here as the events and components are discrete and finite. Perhaps statistical theory is also needed. The first step is to carry out more detailed analysis of TOLG and review all the assumptions and derivations. An overall theoretical model needs to be produced for computer analysis as I believe the TOLG model may be beyond the mathematics available today. The problem may eventually be solved by adopting a form of mathematics based on PDE's or stress tensor calculus. This would produce an analysis similar to that which Einstein used to develop his G.R. theory. Perhaps the maths of General Relativity using Riemann tensors (which seems to produce the best gravity calculations in most cases) may still be appropriate? How ironic that would be.

THE FUTURE

These TOLG problems do not invalidate TOLG far from it, they may help to prove it. And the discussion of proposed "solutions" and the explanations of how TOLG works in some key situations is not a complete proof of TOLG either but it is a start. The question of how to prove TOLG is correct and is the best gravity theory is perhaps still outstanding. Please read on to other chapters for more information and ideas on how TOLG may be validated. I don't want to just hand it all over to the scientific community for a final decision but maybe that is what will happen. I would like to be able to play an major role in ongoing research but that may become very difficult. Anyway at least I have started the TOLG ball rolling.

CHAPTER 6

Three Worlds Comparison

This chapter compares the three gravity theories, or perhaps I should say the two gravity greats and me. It reviews Newtonian empirical gravity, Einstein's relativistic gravity and my light based gravity. It is a bit like Galileo's famous comparison of the world systems, the emerging and correct helio-centric model, the intermediate model and the existing but incorrect earth centred model. But unlike the earth centred universe being simply wrong, none of the earlier gravity theories are completely wrong. They all have something to offer. Therefore I have not used a "Simplico" or metaphorical madman to present any one theory although I am sure there are many that believe I should have used it for my own gravity theory. But I know that idea will soon be completely dispelled as the new theory is finally proven. All three are simply presented on their merits with some background as to their original development and some comments on their strengths and weaknesses.

The three theories of gravity are called the three Ms for the Mystical theory, the Mathematical theory and the Mechanical theory. Of course Newtonian gravity is the mystical theory because of its unknown, unspecified and hence mystical cause of an action at a distance. Newtonian gravity is not wrong, far from it. It still works very well and solves most gravity problems. But it is based on a mysterious attractive force and an unrealistic infinite speed of action. Then naturally enough, G.R. is the mathematical theory of gravity. Similarly, G.R. gravity is not wrong, quite the opposite, perhaps it is too correct or should I say too complex. The mathematics of G.R. seems to solve all known gravity problems and then some, depending on what value is given to the constants. Finally, in more ways than one, there is my mechanical gravity theory, TOLG. It is called a mechanical theory for obvious reasons. This new theory is of course not wrong or should I say what is there is not wrong. While it may have problems and needs more work including development of a full mathematical model, it is the best of all theories. When it is completed it will explain all gravity situations and then some that haven't been explained by other theories. It will finally enable derivation of Newton's gravity constant, G, from other known properties or parameters of physics.

But before I look at these three theories in as much detail as I can muster and as much as I believe is sufficient for this book here is some light background to the laws of gravity.

GRAVITY RULES, OK

Even though the basic law(s) of gravity seems to be well enough understood by most people and also quite comprehensively analysed by modern physics, there are still some areas of interest and challenge about the subject. Not least is the unusual or some say mysterious concept of an "action at a distance". This aspect of gravity is not explained by the two current theories at least not in my mind and perhaps not in the minds of some physicist. This part of the gravity puzzle, action at a distance, greatly challenged Newton. He couldn't explain it to himself or others and he had to accept it as an unquestionable given. It is still a mystery to most people today. By contrast, Einstein tried to solve the main question of gravity by using another concept that is not directly associated with any attractive influence. Instead he created a new concept about changes in the "geometry of space and time". In General Relativity, gravitation is said to be due to a change in space/time around any body with mass (energy?). or something he called spacetime curvature This apparently causes objects to move (accelerate?) towards each other as they travel inertially (no force?) along "new curved straight lines" called geodesics.

Most physicists accept G.R. as the "complete answer" to the mystery of gravity. This theory is brilliant and apparently addresses many other mass/energy/force questions but it is very complicated. And it simply replaces one mystery with what seems to me to be another version of a mysterious action at a distance. Has space/time curvature been more simply explained or related to any more obvious fields or forces by anyone since it was conceived? I don't believe so. The concept of using simple physical properties of matter to explain physical behavior of matter is basic to science but it doesn't seem to apply to G.R. Yes I know there are many "proofs" of G.R. and I have already examined some of them but I am still not convinced it is the real interpretation of what happens. Newton's unexplainable action at a distance simply becomes an unexplainable mysterious warping of space/time at a distance. Both are somewhat impractical and do not use simple explanations or obvious physical properties of nature in describing the process. Most physicists seem to think this is not a problem, especially as G.R. seems to work, but I believe it is a problem. Yes I can hear the purists saying "this guy just doesn't understand G.R." and perhaps there is an element of truth in this. In reply I simply say yes I am a "child" in this area, but as a child it seems like a case of the "emperor's new clothes" to me. When I "grow up" (metaphorically speaking?), my view will be more soundly based but just as objective and I still don't think I will see any "clothes".

There is a type of unwritten rule in science that the real physical meaning or cause of some situation or event must be found or at least identified before a new theory can be accepted. Or in more general terms a new theory must be consistent with the existing framework of real science. This is in addition to another law of science that any theory must be supported by repeatable experiments or measurements. These ideas have been a cornerstone in the progression of real science but have they always been followed or are they always useful? These "rules" have not always been precisely followed and have often been ignored to some extent in order to establish a theory. Some theories are so complex that they are beyond simple comprehension. For others there is often a delay in developing the necessary modelling and mathematics. And sometimes the technology required to investigate the theory is not yet mature. This "rule" is also sometimes challenged as being too restrictive, as in the case of Newton's law of gravity. The mathematics seemed to produce the right answers so the theory was considered correct, even if it didn't it explain everything completely. Does this also apply to G.R.?

It has been suggested that such rigid rules are too restrictive and may prevent new paradigms of science from surfacing. Perhaps it is considered as being too limiting on the development of scientific ideas and is therefore an impediment to the progress of science. It was an important component of physics research in previous times but in recent times it seems to have become less relevant. Physicists and some other scientists have often adopted an "everything is possible" approach, to justify strange new theories and to try and provide answers to mysteries. This has included adopting ideas such as information transfer greater than light speed, challenging the laws of thermodynamics including breaking energy conservation and many more. This has often resulted in the creation of extreme physics theories just to derive a simple and quick fix solution. There is an element of the end, a quick fix and fame, justifying the means, meta-physics. But if the means involves breaking the established laws of physics, then a very good answer is required and the existing theories it damages need to be healed and not just heeled. While it may be a useful developmental tool to investigate alternatives, I believe this approach is potentially dangerous for science. It may let in more mistakes than it keeps out. So how do the gravity theories or the three M's fit into these rules? Well here are the three M's in a gravity shoot-out.

NEWTONS *MYSTERIOUS* GRAVITY

Undoubtedly the most important scientific work into the theory of gravity was carried out by Newton hundreds of years ago. Other theories on gravity that were in vogue about the time of Newton, surprisingly some based on magnetics and some on mechanics such as kinetics and fluid pressure. However they didn't survive and his work stands alone as the greatest of them all. His work on gravity was developed around the time of the Great Fire of London and the famous Black Plague. This later disaster supposedly played a role in his work by forcing him to stay away from Cambridge University and the plague ravaged cities and to work at home on the farm, presumably in his orchard, where he came up with his brilliant theory about gravity.

The background to his ideas may be buried in his papers but it seems as though he wrestled with the need for a force to keep the planets in orbit or at least to make Halley's Comet return. He was also fascinated by what made things (not just apples) heavy and be "attracted" to earth. He may have considered many factors such as the shape of a body, its colour, its size and content whatever that meant. But he soon came to the conclusion that none of these was important, just the weight of a body. But the terms weight, mass, heaviness and many related terms such as force, had similar connotations. So which one was correct? The scientific concepts of atomic matter, atomic mass or atomic content as well as ideas such as atoms, protons and neutrons, where still centuries away. He knew from Galileo and his own experiments that all bodies accelerate the same when they fall to earth regardless of mass. But his formula relating mass and acceleration soon showed that this was so only if the force of gravity was directly related to the mass of a body. So it was obvious that mass in a weight sense was a key factor in any gravity formula. But this was almost too obvious saying that the force of weight due to gravity is related to the weight itself.

This question about the relationship between weight or mass in a gravity sense and mass in a more atomic substance sense still surfaces today. At least Newton saw the need to include the weights or masses of both bodies in the gravity relationship. At that stage of scientific knowledge there was no measure of the mass of the earth but there must have been a good understanding of its volume and density. These would have enabled a reasonably accurate estimate of the weight or mass of the earth to have been made. But what other factors were involved in deriving a gravity formula. The next factor in gravity that was almost obvious from planetary observations was the distance between the bodies. It was not obvious that distance was an important factor from simple earth bound situations as all bodies seemed to weigh the same and accelerate the same regardless of how high up they were. Even for a vertical displacement of many Km, weight loss or a reduction in gravity would not have been detectable but if so that would have added more complexity. Fortunately the known planetary movements and estimated distances proved distance was a key factor. But again what was the relationship between gravity and distance. Logically it had to be an inverse type of relationship but what form should it take. Perhaps Newton had his own ideas but Hooke proposed the inverse square law. There doesn't seem to be any logical reason for this law in Newtons work, just that is seemed to provide the best answer. So finally Newton had the basis for his gravity formula, the product of two masses and the inverse square of the distance between them.

$$F \sim \frac{m_1 * m_2}{d^2}$$

Now all he needed was a constant to make it all work and from his universality rule for gravity, this constant would be the same for all gravity situations everywhere.

$$F = \frac{G * m_1 * m_2}{d^2}$$

The constant G came out of a simple gravity situation, the force or the weight of a unit mass on earth. If the force of 1 unit of mass is the same as its weight by definition, then definition $F = m_1$, and $G = d^2/ m_2$ where m_2 is the mass of the earth and d is the diameter of the earth. Newton probably quickly substituted his estimates of these measures in the formula and came up with his gravitational constant to make it all work. Eureka! Or was that another famous physicist. Newton soon found his formula worked on planetary orbits after he had estimated the mass of the sun and its distance from earth from our own planetary orbit. Then all the other planetary orbits were explained, well almost. He then presented the results of his work in the Principia which became a best seller in his lifetime. But he couldn't sell the rights to Hollywood or even Pinewood Studious as neither existed then.

Newton developed the basic theory of forces and movement of bodies in general and gravity in particular over a few short years. He developed his form of the calculus (fluxions) to solve the types of continuous small step analysis problems which arose. His laws of motion and his basic equations linking force, mass and motion are the core of physics and the basis of almost every branch of physical engineering today. Understanding and using his equations is the source of a lot of examination misery for students of physics. Today, engineers have translated all the formulae and analysis to computer programs to simplify problem solving and unfortunately to reduce their own value. His equations still work well and keep us all safe and happy on earth. We all owe Newton a lot.

While Newton was brilliant by any standards, he was human and had his failings. He was unable to work out what caused this apparent force of attraction between bodies which was the centrepiece to his gravity theory. This was perhaps Newtons biggest failing. He was unable to come to terms with a basic concept or some say rule of science that all things should be explainable in simple physical terms, and not through mystical concepts. He was unable to relate gravity to any other known forces such as electro-magnetism and mechanical forces such as "fluid" pressure. But perhaps these were all minor failings for such a great man. His formula worked and put earth bound gravity calculations and planetary motion studies on a sound mathematical footing at least for a few centuries.

NEWTONS MYSTERY

So, Newton's gravity was a force of attraction caused by some mysterious action at a distance. While the scientific concept of a field had not really emerged by then, perhaps Newton had the idea that gravity was caused by some kind of force field. There was the parallel with magnetic forces of attraction (and repulsion) that were known about at the time, at least in a very limited way. Gravity also had the same effect as a conventional direct external push force. It could and was being explained as being similar to some kind of pressure force, but was this idea known to Newton? It seems not, at least from his main writings or if he knew about it he may not have believed it. Was Newton's unknown cause of gravity a mistake? Did it break the "rule" of science? His use of a mystical force of attraction at any distance, created a problem with the acceptance of his theory in some circles, especially mainland Europe. Some doubted his ideas because they had a mystical component. Some say he didn't see it as a problem, or at least not one that anyone needed to be concerned about. So was it a mistake at all or just a mystery? It did not change the outcomes or calculations of force and movement from his equations in any way. But it did leave his theory incomplete. This is the area that my new theory addresses and demystifies. The new theory solves the mystery of Newton's gravity.

NEWTONS MISTAKE

There is another more significant problem with Newton's theory of gravity. There was no allowance in his theory for the speed at which the force of gravity could work or simply the speed of gravity. Perhaps if he had thought a bit more about the cause of gravity he may have realised what was in fact a greater question, and eventually his greatest mistake, what is the speed of gravity?. In fact the question of the speed of gravity doesn't seem to be very clearly addressed in any of his work. Or did he simply fail to understand the physical implications of an infinite speed.

It appears that he assumed that gravity acted instantaneously, even over infinite distances so it was effectively based on an infinite speed for gravity. Even though he developed a complex form of calculus to solve the difficult maths of gravity he doesn't seem to make any allowance for the concept of time in his theory. If he had realised that applying a force instantly over infinite distance was not logical and a realistic speed limit had to apply he may have seen the light so to speak. It is possible that he knew it could not be infinite, but perhaps thought that it was so very fast that it didn't need to be addressed. He would have been unable to relate it to anything else such as the speed of light. This is excusable because at that time the science and theory of light was in early stages of development. Even the concept of a fixed and finite but very fast speed of light was yet to be fully accepted. A realistic estimate, let alone measurement or calculation of light speed was still a long way off.

While his equation for the force of gravity solved most known problems of earth bound motion as well planetary motion, it has some minor imperfections, especially in extreme situations. His studies and his theory did not solve the major question, "what is the cause of gravity?" This mysterious problem of what is the attractive force of action at a distance seems to have caused Newton considerable concern. This unsolved problem may have been what really led him away from the more analytical studies in the areas of physics he previously enjoyed so much, such as optics and forces, and drove him to spend considerable time and effort studying the less practical "sciences" such as alchemy and also other mystical areas and religion. Was he perhaps looking for a solution to the mystery of an action at a distance through these strange sciences?

So, was Newton wrong? Obviously not! Well not in general terms that is. His theory of gravity was not really a theory of what gravity was or what caused it, it was simply a heuristically derived mathematical explanation of how it works. Or more precisely, what gravity does to bodies that have mass and get close to each other but don't move much. Newtons missing component or Newtons Mystery was that he could not find the cause of gravity. However, more significant than this was his view that gravity acted instantly. Hence, his calculations made no allowance for any speed of gravity and his law was purely a static one. Was this really Newton's only significant mistake? Yes it seems so but it eventually led to the downfall or at least the need for a replacement to his theory.

Given the overall enormity and value of his work, the mysteries and mistakes of his theory are perhaps trivial. The fact that his equations were good enough to finally prove the earth centred model was correct, to help find some missing planets and also to get man to the moon, made these problems insignificant at the time. They still are today. That is at least until now with my new theory of gravity addressing both of these problems. Have I solved Newton's mystery and corrected Newton's mistake. I believe I have.

The speed of gravity became a very interesting question following Newton. His theory worked well in most situations and this added to the view at the time that the speed of gravity, if not infinite as Newtons laws assumed, was indeed very fast. But if it was not infinite what was it. This issue has also raised some very interesting challenges for all new theories of gravity. There were contributions to this question by greats such as Pascal the French mathematician who estimated that the speed of gravity had to be very fast to avoid aberration which would occur if one or both of the bodies were moving very rapidly. There is a strong view still held today by some researching gravity, that the speed of gravity is very much faster than the speed of light. However, the speed of gravity was eventually related to the speed of light and in current gravity theories and in TOLG it is the same as the speed of light.

EINSTEINS *MATHEMATICAL* GRAVITY

The next major work on gravity was by the most famous scientist of the twentieth century, Albert Einstein. His work on gravity or more correctly his work on general relativity followed his famous work on special relativity. It was perhaps part of his aim to relate and connect all the parameters, fields and forces of nature as they are called, in both inertial and accelerating situations. He had determined that speed changes time and space (length) and perhaps he wanted to determine if other factors such as acceleration or mass changed them also. To unify all the effects and theories of physics was considered a great pursuit of physicists at the time and still is today. Perhaps he also saw the need to include the speed of gravity in a new theory. This aspect of Newton's gravity required correcting because an infinite speed was not logical or did not seem to be consistent with Special Relativity. But what was the speed of gravity and how could it be included in a new theory. Was Einstein also driven to find a real cause for gravity? Perhaps finding the cause would help to find its speed. This would then produce a complete theory which would fill the gaps in Newton's theory and link all the forces of nature? If only!

Einstein was obviously aware of the situations that Newton's theory did not explain, especially the unusual orbit of Mercury. It seemed highly likely he believed this and similar errors were due to the fact that gravity had a finite speed and hence a finite time of action over a distance. However, what was that speed and what caused it to be finite and what were the other implications of such a finite speed? Other physicists had toyed with the idea of light speed for gravity and how it could be factored into Newtonian gravity. Some physicists had already proposed that if gravity acted at light speed this alone would explain Mercury's perihelion advance. Einstein must have been was aware of this work as it was done by some of his colleagues.

In any case, whatever the background was, he eventually came up with an idea about accelerating reference frames and a new way of redefining space/time. An idea based on a new geometry and a new mathematics for describing this geometry called Riemann geometry. He based his new theory of relativity or gravity on this new curved space/time represented using tensors. He used the Riemann Stress Tensor and the Ricci Flow Tensor. And after considerable effort he produced his bold new theory which shook the world and it is still shaking. His new "gravity formula" can be presented as;

$$R_{ab} - (g^{ab} R / 2 + g^{ab} \Lambda) = 8\pi G * T_{ab}$$

While this may seem a simple formula, it should be noted that the **R** and **T** are Tensors or arrays of partial differential equations and other factors are metrics or other relations. In fact it is very complicated and even as Einstein himself said, extremely difficult to work with. It is interesting that Newtons gravity constant, **G**, is included in the equations and they also include the speed of light. It has supposedly been proven to be the correct theory of gravity but perhaps this was more due to its mathematical success than to it being the correct interpretation of gravity. The general idea is complicated and is still under review and still has critics and doubters. I still have problems with the more esoteric aspects of it, especially space/time curvature at a distance from mass/energy, which is the cornerstone of the theory. The mathematics of G.R. is more complex than Newton's simple law but reduces to it in static or slow moving circumstances. And yes, G.R. solved most of the special gravity cases that Newton's law couldn't In particular it solved Mercury's orbit problem simply because it has a speed of action for gravity which is the speed of light.

However, even given it success, perhaps there aren't that many physicists, let alone many ordinary people (if any?) who fully understand this complex mathematical and meta-physical explanation. Few can say without any doubt in their mind that they understand why "space/time curvature" actually happens. It seems to me to be another version of a mysterious action at a distance. Has it been more simply explained by anyone since then? I don't believe so. Yes I know there are many "proofs" of this warping of space time and I will examine some of them later but I am still not convinced it is a satisfactory explanation. Well there go some more readers. And therein lies what I believe to be the enigma in Einstein's work. How did he arrive at the speed of gravity in his theory and his formula? Was it an input or did it just come out of the new mathematics?

EINSTEINS ENIGMA

How did Einstein arrive at the "fact" that the speed of gravity is the speed of light? Einstein may have believed that gravity could not have "travelled" faster than the speed of light as some of his peers seemed to believe. This would have contravened Special Relativity which states that the speed of information transfer could never be greater than the speed of light. And surely gravity or the "gravity field" carried information about the mass and location of the source. It was also obvious from all experiences at that time that the speed of gravity was very fast. So did he arrive at this speed by assuming gravity acted as fast as possible but couldn't travel faster than the speed of light. Or did he simply think that somehow space/time curvature was related to some kind of E/M field and it was already known that the speed for E/M fields was the speed of light. Did he then set out with this speed component as a starting point or basic assumption and then derive his equations to fit or did the speed of light automatically come out of the mathematics he used to formulate his theory just like it did for Maxwell? I will leave that one for the G.R experts to answer. This seems to raise a major question about G.R and while there may be obvious answers in the theory or in his papers, this still seems a mystery or should I say an enigma, at least to me. And did this theory completely solve Newton's mystery as to the cause of gravity in everyone's mind.

While the speed of gravity must have been one of the driving forces behind Einstein's work on General Relativity, perhaps the other was the need to fill in the other gap in Newtons work which was "what causes gravity" or what makes bodies (mass) seem to be attracted to each other. What was the mysterious cause of this action at a distance? Einstein thought he had the answer with his new four dimensional space/time continuum that changed or curved due to the presence of mass or energy. He challenged the conventional wisdom of space being based on linear Euclidean geometry. Parallels with an earth based spherical geometry and geodesics or curved "straight lines" were perhaps a starting point. However, to me at least and perhaps to others, it seems as though Einstein replaced Newtons unexplainable action at a distance with a similarly unexplainable space/time curvature or warping at a distance. Both are somewhat mystical and neither uses any simple measurable physical properties to describe the process. That is, they both broke the rule and lack a plausible explanation as to the real cause of the force of gravity between two bodies. Of course one day such a simple and plausible explanation that better explains space/time curvature may come along. Perhaps this new gravity theory may be part of that explanation and enable G.R. or at least its mathematics to be reborn, but based on a better explanation.

EINSTEINS ERROR

So what was Einstein's Error? Well again given the enormity of his work and his many contributions to science, it seems a futile waste of time to even consider such a question. However, I had to research some aspects of his theory in order to find a reason for the speed assumption and for other factors that lead him to his "solution". In particular I carefully considered the cornerstone of G.R. and that is the equivalence principle. Perhaps this was an area of potential error. Einstein Equivalence Principal (EEP) basically states that force and acceleration are the same thing. To some extent this is not a new idea. Newton showed this in one of his famous equations which related force to acceleration with the "constant" being mass. For any given mass, force equals acceleration in Newtonian physics. But Newton's "equivalence" was causal. How did Einstein reinterpret this? The famous thought experiments of the falling man (or was it a real event?) not feeling anything and the person in a lift or spaceship experiencing either gravity or acceleration but not being able to differentiate between them, are the basis for EEP. He came to the conclusion that force and acceleration are more than equal, they are the same thing or equivalent. In particular he saw no causal relationship between them. One didn't come first and the other followed, they happened at the same time. Shades of Newtons instant gravity!

However, it is now widely understood that this equivalence principle is not completely correct at least in one respect. There is, in fact, a small but detectable difference in any gravitational field strength over space due to the effects of radial distance from the source and the conical shape of the so called "gravity field". It is now generally called the weak equivalence principle (WEEP) because of these errors. Perhaps this interpretation would have made Einstein cry also. The question is how, if at all, does this change the situation? How does it change G.R. and its application to gravity? The current interpretation is that the W in EEP has no significant effect but I wonder? Similar, apparently infinitesimal discrepancies have often led to the complete overthrow of a theory. And what about causality, how is it interpreted in G.R. If gravity has a finite speed of action isn't that proof of causality?

Another problem with G.R. is the way it can't be applied to very small scale situations, especially to subatomic particles. Other subatomic theories and forces have been developed to overcome this problem with the so called strong and weak nuclear forces coming into play at these distances. Is this an error that makes G.R. questionable? Does the basis for and mathematics of G.R. indicate that "small" scale application will be a problem? Or does it just come out of applying G.R. to such small scale situations and finding it doesn't work? Why isn't G.R universal and why doesn't it apply to every scale? Is it simply quantum mechanics or a Plank scale issue? Also how does G.R. relate to these other subatomic forces? It would seem that there should be some way to tie these forces and fields all together. Einstein was unable to do this with G.R. and space time curvature. These questions are currently beyond my understanding of G.R. but would be interesting subjects to pursue. But were there any other errors in his work?

Perhaps his major error was in assuming, or is it proving, that gravity bends light and then accepting at face value the Arthur Eddington light bending experiment during an eclipse of the sun. This experiment measured star displacement near the sun during an eclipse when it was possible to see stars close to the sun. It was then used to "confirm" this prediction of his theory and to confirm his theory in general. Everyone assumed that it was only the suns gravity that bends light near the sun. Does it really happen? Many subsequent measurements have confirmed that there is light bending near the sun. But are there other possible explanations? I believe there is a simple one based on light speed changes due to changes in the media parameters. It is well known that the sun has an atmosphere of hot ionised gas or plasma which extends many millions of Km into space. This plasma would also affect the properties of space such as the electrical permittivity ε and the magnetic permeability μ. Both of these or at least ε would be changed and this would slow light down. And as we all know, slowing light down bends it. This is the basis of diffraction and this is discussed in the chapter on light. Therefore light near the sun would slow down and bend, perhaps more that enough for the measured bending supposedly due to G.R. So is the assumption of light bending by gravity wrong? I believe it is as there is a simpler answer and I am surprised no one seems to have considered this explanation before. So was this his major error? Perhaps his famous expression that goes something like "if it (light bending due to gravity) didn't happen then I would be sorry for God" may come back to haunt his memory.

There are supposedly many other "proofs" that G.R. is correct but I won't go into any of them here. Many depend on gravity bending or effecting light and perhaps these may need other explanations. Again the interpretation of some experimental results may have been too quickly "blamed" on G.R. and there may be other simpler explanations for them. Although most physicists say the theory of General Relativity is correct, it is still under review and still has some problems that need solving. There are question about its completeness and why it has a (variable) constant called lambda and also why it still uses Newtons gravity constant G instead of more fundamental physical properties of matter. The new theory of gravity described in this book should shed some interesting light (so to speak) on G.R. and the gravity solution it offers. The new theory proposes a very simple explanation for gravity using well known physical properties of nature. It does not require any type of mystical force field or strange curvature at a distance and it also challenges some other aspects of G.R. In particular it doesn't need Newton's gravity constant G as an input. This factor and the speed of gravity are simply derived.

TOLG *MECHANICAL* GRAVITY

The final theory of gravity to be discussed is The Theory of Light Gravity, TOLG. This new theory was based on my need for a realistic or perhaps I should say an engineer's explanation for gravity. I believe that gravity must have a more mechanical explanation. I had a number of ideas but in particular I thought that if gravity works at the speed of light then real light must be key. I also realised that gravity was a causal force. That is, force comes first then mass experiences the effects of that force. There is a delay in action just as there is in any other causal situation and this delay is due to the speed of gravity. I didn't think it could be due to a mystical attractive force but was most likely caused by a push force. Then I thought about what could cause this push force and realised it was most likely due to some form of pressure difference. But how could a pressure difference be created? Then I had a brainwave (if that is the right term) and realised that it was caused by shadowing of some kind. Shortly after this I did some research and found that pressure gravity theories were old hat. Someone had beaten me to the punch. But why hadn't such theories gone anywhere? Why did they not succeed? More research soon showed why. They had all created some form of conventional fluid to cause the pressure. Sometimes this was related to the light carrying ether sometimes to some other mystical type of particle but none could be specifically defined and none could be found. So if shadowing was correct, what light was causing the shadow?

After some simple calculations it was obvious that light as we know it would not work. What was required to complete the picture was the right kind of light. If the pressure due to a kinetic effect from light was to be high enough, a new high momentum light was required. It needed to be in the form of a high energy ultra short wavelength background radiation that could not easily be detected. Shades of the old undetected ether story but I wasn't deterred by thoughts of failed MM experiments. This was a job for super-light. But where was he or she? Then I had my next great idea. I found a possible source for this type of light in my new atomic model. Then I realised that the whole mystery of gravity, its cause and speed and everything was uncovered. No more mysteries, no warping, no strange constants, all simple physics.

My new theory of gravity needed to have the same results as Newtonian gravity for most simple or static situations and the same solutions as G.R. for relativistic situations. The first requirement is obviously satisfied. TOLG shows how the force of gravity is directly related to the mass in terms of atomic matter of each body and it shows how the inverse square law is derived. It also shows how Newtonian gravity is correct for static situations but will start to produce errors as the speed of either body increases. As for G.R., I am still working on the connection and it will need the final mathematical model of TOLG for any real comparison. However TOLG is caused by light so it obviously works at the speed of light which is the same as the speed of gravity in G.R. so they are the same in this regard. As for other parallels or comparisons, time will tell but it seems to me that TOLG will work out to be very similar to G.R. in many ways, apart from curvature. One interesting observation is that it immediately solves the age old question as to why all bodies fall at the same rate. All bodies are made of the same atomic material (protons, neutrons) and it is these that fall under gravity so it is obvious that all bodies will suffer the same gravity effects. TOLG doesn't really care if it is putting pressure on one particle or billions of them, it is all the same. But this atomic picture has other interesting consequences which are addressed later.

TOLG is a non standard theory or an alternative theory, and this automatically puts it under a cloud of suspicion. That is the least of my concerns. What really concerns me is the difficulty I will have in proving it from my non scientific position. It will need to be proven piece by piece. Perhaps the first task is to prove that PP exist and then to demonstrate it and then show CGBR exists and then to measure it. This will be challenging enough as PP hasn't even been proposed before by all the physicists that have ever lived, let alone observed or measured with all the myriad of expensive equipment available today. Then a process for deriving a mathematical formula for gravity based on this new theory is required. In other words, Newton's law of gravity with a speed of light component needs to be derived or more likely a matrix of differential equations will be required similar to Einstein's G.R. The mathematics will then enable solutions to gravity problems to be formally calculated according to TOLG. Wow, have I got some work to do. Perhaps I may be able to obtain assistance from the broader scientific community, especially from gravity physicists who may also be eager to progress this analysis and share the glory. Any contributions toward finalising my theory and clarifying the basic ideas it is based on, such as PP light, could be helpful.

It is important to note that TOLG does not completely disprove existing theories; in fact it kind of proves them. What it does is extends them in elegant and scientifically sound way. Therefore this new theory doesn't change gravity calculations in general, and is also unlikely to cause any significant change in the application of gravity theory to most everyday gravity solutions. The planets will continue to orbit the sun and it won't really have any immediate impact on any aspects of our daily lives. But it is profound nonetheless and it definitely has the potential to make major changes to many areas of scientific research, especially in gravity related fields. So is it all roses or are there some problems with TOLG? Well yes there are, even if I don't like to say so myself. Perhaps the most significant ones are associated with the light which creates TOLG. How can it be identified and how can it be detected. Then what are the collision and deflection process that cause shadowing and what about problems like shielding. I have attempted to identify atomic based light reflection mechanisms for CGBR in the section on the atom and on light but more work is required, mostly theoretical and mathematical.

One of the major problems with shadow gravity is shielding. This is discussed in the chapter which described TOLG and further in the chapter on TOLG problems. I believe the problem is only a minor one at best and may in fact have some interesting outcomes. It will need to be addressed in the complete mathematical model of TOLG and this should quantify the effects. I am confident that it will be found to be just part of the process of scattering and randomising CGBR again after collisions and will therefore have no impact. Another potential problem is how to define the reflection or capture process that happens during atomic collisions. What is the mechanism involved and how is this connected to the process of pressure which creates gravity. And what is the source of CGBR? How and where is it created in the cosmos and what are all the energy equations for it? Well as I have said before there is much work to be done but I feel very confident the solutions to these problems will only confirm the theory.

Are there are immediate advantages or benefits from TOLG. The reader might ask whether my new theory predicts anything new about gravity or proposes any useful new outcomes in gravity physics. Perhaps it will produce beneficial outcomes and help create new and hopefully non destructive scientific endeavours. I hope the first outcome is not a new war machine like it was for nuclear research. Here we go again, military applications come first, but they have all the $$$$. Does TOLG show how to control of gravity or enable an anti gravity device to be made? Well sad to say, nothing like this seems possible from my new theory. In fact it really supports what we generally know about gravity and that is it will most likely be impossible to change. But TOLG definitely improves our understanding of gravity by addressing the question about the source of the "gravity field" so to speak. But it confirms that the control of gravity is most likely well beyond us if it can be done at all. Further research will provide a better understanding of gravity but TOLG definitely won't make big money in the short term. Gravity calculations won't suddenly change or be easier and unfortunately our weight won't change, which is more of a loss for me as the weight control industry is a potential gold mine. This lack of any potential short term gain, especially any financial gain, from applying my new theory will perhaps limit its press coverage. There won't be any immediate new developments associated with gravity as we know it, but there may be some new and interesting promotional opportunities. I have already mentioned Hollywood, say no more. Of course the search for potential benefits using my new theory will be a fascinating area for future work and many will climb aboard.

Most experts will undoubtedly say that this theory or at least the analysis of it is not very scientific. I accept that the depth of analysis is not what is normally given by researchers in large scientific establishments in support of new idea in physics. But the lack of detailed analysis and mathematics should not prevent it from being considered in a scientific way. There is more than enough information on the basics of the theory and the ideas behind it for more in depth analysis to continue. I will certainly continue with my own research and analysis and only hope that other experienced physicist with access to the necessary extensive resources take up the challenge. Perhaps my bold new theory will easily be cut down to size by reference to some simple law of physics or more likely to some extremely complicated aspect of QCD theory. But I don't believe this will happen, at least not as quickly as some may believe. I can only hope that proton photons (PP) are soon discovered from the work at CERN or some other research establishment or perhaps even by mathematical theory. I will certainly continue to investigate ways that PP may be created or detected or better related to the existing atomic model.

CONCLUSION

Well who in your view won the shootout? Was the comparison fair and did I identify and discuss all the problems, benefits, advantages, strengths and weaknesses of the three contenders. I have attempted to summarise them in the attached table. Should I have included more theories in this comparison such as scalar tensor field theories or what? Did TOLG win in your mind or did I undersell it. Perhaps I am a better physicist than salesman. In the meantime please read on and try and find out more about the impacts of TOLG. Try and get a feeling for the potential benefits this new theory may introduce in due course. Who knows, perhaps you can also contribute to the ongoing development of this bold new challenging theory of gravity and come up with some new ideas on how it can be proven. Or perhaps you can put me out of my misery by proving it can't work. Now there's a challenge but I don't think it will be successfully taken up! Perhaps just finding all my mysteries, enigmas, mistakes and errors will take time.

COMPARISON OF GRAVITY SYSTEMS—3Ms

SYSTEM	NEWTON GRAVITY	EINSTEIN GRAVITY	TOLG
BACKGROUND	MYSTERIOUS	MATHEMATICAL	MECHANICAL
SOURCE	ORBIT MECHANICS BEST FIT LAW	GENERAL RELATIVITY RIENMANN MATHS	KNOWN PHYSICS LIGHT DRIVEN
CAUSE	UNKNOWN CAUSE ATTRACTIVE FORCE	CURVATURE (???) OF SPACE/TIME	LIGHT PRESSURE SHADOWING
MAIN FEATURE	SIMPLE LAW WELL KNOWN	NEW GEOMETRY COMPLEX MATHS	NEW LIGHT NO CONSTANT
WEAKNESSES	UNKNOWN FORCE INFINITE SPEED	CONSTANTS "G", "Λ" WEEP, BENDS LIGHT?	NO LIGHT YET NO FORMULA YET
STRENGHTS	BY A GENIUS SIMPLE MATHS	BY A GENIUS SIMPLY BRILLIANT	COL PARKES???? SIMPLE PHYSICS
ADVANTAGES	SIMPLE FORMULA MOST SITUATIONS	GENERAL FORMULA MANY SITUATIONS	SIMPLE CONCEPT ALL SITUATIONS
DISADVANTAGES	STATIC SOLUTIONS NO GRAVITY SPEED CONSTANT "G"	TOO COMPLEX UNDERSTANDING? CONSTANTS VARY	NON STANDARD NOT YET PROVEN TOO NEW/SIMPLE
LIMITATIONS	CAN'T HANDLE MOVEMENT	CAN'T HANDLE DISCONTINUITIES	CAN'T HANDLE SHEILDING YET)

CHAPTER 7

Atomic Implications

There are two extremes of physics that seem to be the major focus of most scientific attention and research activity these days. Large scale physics for the cosmos so well pictured by the "travels" of the late Carl Sagan is one of these. The implications of TOLG on the cosmos and his travels are addressed in the next chapter. The other extreme is small scale physics where physicists are investigating the atomic model and subatomic particles. I say investigating because it is like police work. Suspects are identified and evidence is gathered from natural observations, experimental results and blackboard work. This evidence in conjunction with a certain amount of speculation is used to try and build a picture of reality or to "frame the most likely suspect". In most areas of physics, just as in police investigations, a very good imagination is required.

This chapter addresses the small scale end of physics, the atom or more precisely the atomic model. The current standard atomic model is presented and key aspects are discussed. Some aspects of the standard atomic model that may be relevant to theories of gravity in general and TOLG, or light pressure based gravity, in particular are identified. This review may be limited but hopefully it serves the general purpose of relating TOLG to atomic physics. It does however identify some possible properties or parameters of the current model that don't seem to be consistent with TOLG and that may need resolving in some way. To overcome such anomalies possible changes to the atomic model suggested. Again I can hear the experts say who am I to make such proposals but nevertheless I have and I will continue and try to work out what TOLG means for all areas of physics, especially at the atomic scale. Who knows, it may eventually lead to acceptance of a new atomic model.

So let's start a journey through the world of the atom, at least up to my current understanding of it. The history of the atom or should I say more generally the development of ideas about the small end of physics is long and convoluted and what better place to start than back in Greek mythology again. I say mythology and not science as most Greek philosophy was more mythological than real. But at least they put some thinking into the problem and tried to understand things in a more structured way.

THE GREEK ATOM

The Greeks seem to have first thought up the idea of a very small and indestructible building block of nature thousands of years ago. The term atom supposedly comes from a Greek term for indivisible or un-cuttable. However the general Greek view of the composition of all the matter in nature was based on a very simplistic idea of their being only four basic elements, fire, water, earth and air. Everything in nature was said to be composed of these four elements and presumably each of them was made up of "Greek atoms". But were the atoms all of the same type for all elements or did each element in the Greek model of matter have its own atom? This idea didn't go far in terms of a sound scientific theory but it went far in terms of lifetime. It seemed to last well into the renaissance period at least as far as western science was concerned just like the Greek Ptolemy model of the universe. Ages of darkness again.

Then along came the renaissance and some natural philosophers who challenged the status quo and who started questioning the real nature of matter. Perhaps they were influenced by the emerging "science" of alchemy, a mystical half science trying to create gold from lead using a form of "chymistri". However, as history tells us this didn't go very far in the way of wealth creation but it did stimulate a lot of interest in test tube experiments. Many an early laboratory probably owes its existence to such "science". Perhaps one of the main movers in the development of microscopic physics and eventually atomic physics was once again the technology of optics, not the telescope but its cousin the microscope. This device allowed mankind to take a closer look at nature in more detail than ever before. What was observed must have clearly created a strong fascination with the micro end of science. Perhaps the studies into light and the proposal of light particles by Newton and the extensive work on gas behaviour by Boyle and many others, also set minds working on the real micro nature of matter in and around us.

In fact it was probably a combination of all of these developments which created the need for better models of matter on a small scale. A new team of scientist was on the move again to find out more about nature and the atom was reborn after many years of lying idle in Greek mythology. This work eventually led to the development of realistic and reliable early atomic theories and models. Some of the key players and steps in discovering the physics of the atom are listed below.

DIFFERENT ATOMS—DALTON (1766-1844)

There are many great names to mention in this evolution of our understanding of nature at the atomic scale but one of the greatest was Thomas Dalton. He is sometimes called the father of modern atomic chemistry. He worked on many aspects of the basic building blocks of matter, including the atom and its physical properties. He proposed that while there were different types of atoms, each element had its own atom with its own atomic properties and that all atoms of a specific element are identical. He also proposed a theory of chemical composition at atomic level by which atoms combined in certain ways to make other substances. He proposed an idea of atomic weight for each element which effectively gave a comparative weight for each atom. He gave Hydrogen the weight of one as it was the lightest element then known and went on from there to give weights to other known elements. This work was a major breakthrough in categorising known elements and it also helped with the discovery of new elements and compounds.

THE PERIODIC TABLE—MENDELEEV (1834-1907)

Another great leap forward in understanding the atom came from a physicist called Dmitri Mendeleev. He believed that all atoms were somehow related or connected and that the properties of elements gave an indication of this relationship. He is reported as using playing cards marked with each elements atomic weight and properties to try and find a pattern and eventually came up with a basic table of the known elements. This showed the periodic nature of elements and how they were related. His new table with atomic weights and properties was slowly accepted and used to formally categorise all the known elements. It proved to be a major breakthrough in the understanding of matter and showed up some gaps in the table between known elements. This then started a race to fill the gaps and find more elements. The periodic table also started a debate about atomic structure. What properties of an atom determined which position it occupied in the table? Was the atom divisible and if so what were atoms made of? And what did the atom look like and what was its size? Was it like a round hard ball of matter or energy or was it less well defined? By now physicist had come to the conclusion that all atoms were about the same size, but had different masses. What did this mean for its internal structure? What was inside the atom to make it different for each element? These questions were the focus of research from that point on and are still fundamental questions of atomic physics.

THE ELECTRON—J.J. THOMPSON (1897)

Technology was on the move again and gas discharge tubes or vacuum tubes with electrodes and increasingly higher voltages showed some surprising results. This led to the proposal of a small charged, dynamic particle which was eventually called the electron. The discovery of the electron by the English physicist J.J Thompson was a major leap forward. It is the oldest "living member" of the sub-atomic particle family and it is also the only one still standing so to speak. It still has basically the same properties that were originally assigned to it when it was first discovered unlike most other atomic particles that keep being subdivided. The electron is a small particle even by atomic scales and was given a negative polarity from the convention already adopted by Faraday and others due to its behaviour in electric and magnetic fields. Properties such as charge and mass were eventually measured by experiment. This work also helped develop the relationship between electrons and light.

There is one property of the electron that seems to vary and that is its energy. Is the difference purely due to speed differences as in kinetic energy for example or is there some other property such as mass that needs to be taken into account. It is well known that when electrons are orbiting the atom in a captured state they have different energy levels. If they are travelling at light speed, as seems the case, then they effectively must have different masses. How an electron changes its mass and can effectively create different wavelengths of light in energy jump transitions is still an interesting question at least for me. Perhaps there are extensive quantum mathematical explanations for this behaviour but they are beyond the scope of this book and beyond my limited theoretical physics knowledge. Today our knowledge or should I say the model of the electron is quite well developed. The electron is one of a class of subatomic particles called Leptons. It has a negative electric charge of about 1.6×10^{-19} coulombs, and a mass of about 10^{-30} kg which is approximately 2000 times smaller than a proton.

THE "PLUM PUDDING" ATOM—J.J. THOMPSON (1904)

While one of the basic components of the atom, the electron was now better understood, the model of what an atom looked like did not seem to progress much from the old idea of a ball of "matter". The accepted model of the atom around the start of the 20th century was the one proposed by J.J Thompson after he discovered the electron. It is called the "Plum Pudding" model. It had all the components of the atom, whatever they were, in a simple round shaped ball with the electrons scattered throughout like plums or raisins just like an English plum pudding and so the name stuck. What the other matter in the "Plum Pudding" atom was, had not yet been determined, apart from some general ideas about the need for mass and also charge balance to make it neutral.

There were other scientific discoveries around this time that directly or indirectly contributed to the development of the model of the atom. They included the discovery of radioactivity which Thompson was also connected with. It was found that radioactive substances produced particles and changed into new elements. Three types of particle were identified and named, the heavy and positively charged Alpha particle which was later found to be Helium nuclei, the lighter and negatively charged Beta particle which is an electron and the Gamma particle which is a high energy photon. These particles were found to radiate from many elements but the radiation was much higher from particular elements later called radioactive elements. These natural radioactive sources of particles were then used as projectiles in further experiments to determine atomic structure.

ATOMIC NUCLEUS—RUTHERFORD (1909)

Along came Ernest Rutherford a New Zealand physicist who was very interested in radioactivity. He discovered the concept of half life for radioactive elements which went on to be of great use as an aging method, especially in geology. Then he found that Alpha particles were positively charged Helium atoms or more precisely Helium ions or just Helium nuclei. He was fascinated by radioactive particle experiments and he liked to "shoot" radioactivity rays or particles at matter, especially gold foil targets to see what happened. Perhaps he was just trying to see if he could break gold down into other elements and even create lead from gold, in reverse of the alchemists of past ages.

To his surprise some Alpha particles were observed to be deflected in a large way from a thin gold foil target. That surprising result lead to the now famous expression he is said to have made that went something like "it was just like having a cannon ball bounce of tissue paper". Rutherford showed that his results seemed inconsistent with the plum pudding model of the atom which he thought was as dense as tissue paper. Such high angle scattering could only mean the atom must have a smaller, more solid positively charged core. He then proposed a ratio between the core and the estimated overall atom size of around 8000. This just shows how many deflections his assistants had to observe in order to get acceptable errors. Who said physics experiments were easy! The term Rutherford scattering comes from these types of experiments. It is perhaps better known by science students today as the rubber sheet experiment. Many have counted deflection angles of rolling balls in this experiment.

This experimental result produced a new atomic model with a very small central nucleus surrounded by a lot of free space which somehow contained the electrons. This model had supposedly also been proposed at a similar time by a Japanese physicist Nagaoko. There was much discussion about this new model and questions were raised about how the electrons "behaved" and also how the core may be structured. Physicists were seeking a better understanding of atomic structure and started searching for other components. What were the components of the new small nucleus and how did they fit together. These became the big questions of physics in the early part of last century and led to a whole new industry of atom smashing.

QUANTISATION—PLANK (EARLY 1900's)

Plank was a famous scientist who was primarily interested in thermodynamics and black body radiation. It is interesting to note that I was born the year Plank died, which is a bit like Newtons birth at the time of Galileo's death although of course I am no Newton. While Plank made no direct inputs to the better understanding of the atom, his work was important for progressing related ideas. His theory on atomic heat and the solution of the UV catastrophe was a major step forward. He proposed the famous energy equation relating radiation energy to radiation frequency and developed the idea of quantised atomic heat energy which improved the understanding of electron energy and light. This then led to the theory of quantised light and electron "orbits" that was a key step forward in the further development of the atomic model.

PHOTO-ELECTIC EFFECT—EINSTEIN (1905)

Einstein was a friend of Plank and built on his work. Through his photoelectric effect paper he set the scene for the development of a more comprehensive theory of light and atomic or at least electron activity. It showed how light photons required a particular wavelength or a particular energy to interact with atoms or at least with electrons around the atom. The energy level required for this interaction was different for each metal or element. This theory went a long way to developing the fixed energy states and associated electron orbits in a new model of the atom and finally confirmed quantised light. It is also what eventually won him the Nobel Prize in Physics.

PROTONS—RUTHERFORD (1918)

The proton was proposed soon after the concept of an atomic core was proposed. It was needed to balance the electron charge as the atom was known to be electrically neutral. The real discovery again came from Rutherford scattering, this time by firing alpha particles at nitrogen. This generated positive particles that behaved like hydrogen nuclei. They were found to have an equivalent charge to the electron in terms of magnitude but it was the opposite polarity. The new particle was also "given" a much larger mass than the electron. This was needed to make up the overall atomic mass of elements, especially Hydrogen. Again all protons were believed to be identical and indivisible until relatively recently. They are now thought to be decomposable and some even think they are capable of spontaneous decomposition into something else. This area of physics is still under much analysis and discussion and TOLG may be an input to this debate. The electron is still supposedly indivisible but it is now proposed that the proton is made up of three quarks, one of which supposedly carries the charge. But how does this affect proton behaviour? It was also proposed that protons decay but subsequent analysis has shown this to be virtually impossible. Perhaps it is also nor subdivisible after all.

PLANETARY MODEL—NIELS BOHR (EARLY 1920's)

The well known planetary model of the atom with a core nucleus of protons and neutrons together with orbiting electrons was really developed by Bohr in the early part of last century. However his initial orbiting model of the atom had stability and energy problems. No suitable explanation of how the electrons stayed in orbit, without acquiring or releasing energy could be found. In addition the new photoelectric process of photon creation or capture could also not be explained satisfactorily.

THE ORIGINAL "PLANETARY" ATOMIC MODEL
(NOT TO SCALE)

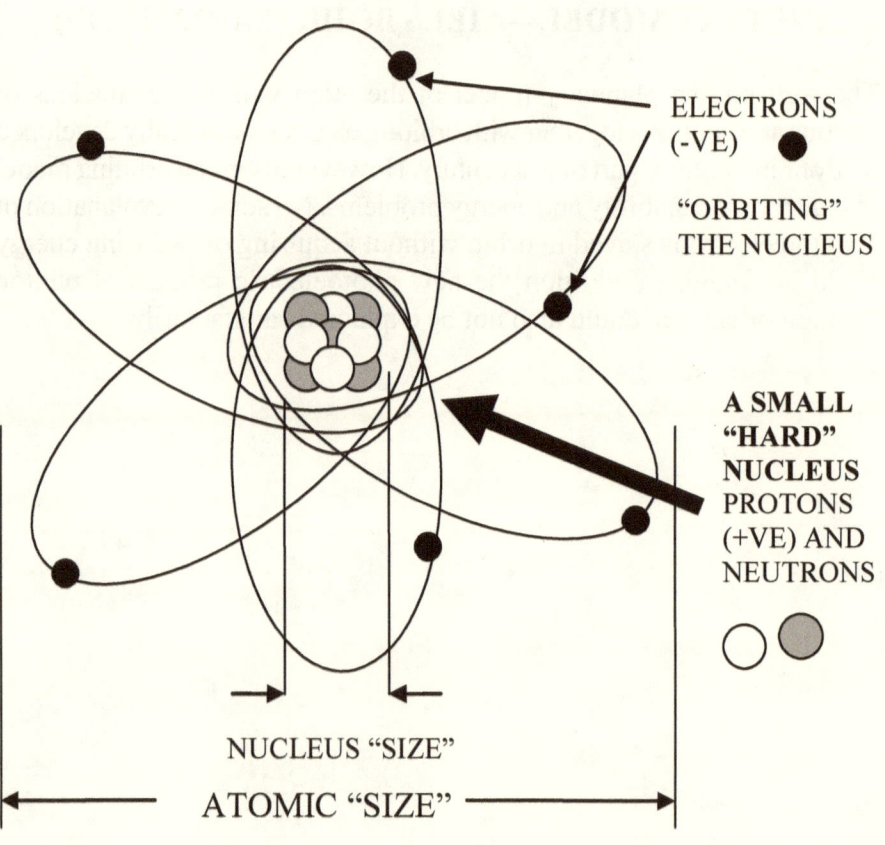

ELECTRONS
(-VE)

"ORBITING"
THE NUCLEUS

A SMALL
"HARD"
NUCLEUS
PROTONS
(+VE) AND
NEUTRONS

NUCLEUS "SIZE"

ATOMIC "SIZE"

HARMONIC ORBITS—DEBROGLIE (1924)

About this time in the progress of physics, quantum theory was being developed and new mathematical ideas were also being used to further explain atomic structure. The quantum mathematical model of the atom slowly took shape and the idea of using wavelengths to determine orbits was a major breakthrough. This amazing proposal was made by French physicist Louis de Broglie. He proposed that all particles, not just photons, but also electrons and all other matter can be considered as having dual characteristics, that is, wave like or particle like. He proposed that electron orbits (circumferences) should be harmonically related to the energy and hence wavelength of the electron. He derived the wavelength from the mass and energy of an electron from Planks equation and hence derived the orbits. All orbital calculations were found to match existing predictions of electron behaviour and his proposal became a hit.

EXCLUSION—PAULI (1930)

While the de Broglie proposal solved many of the problems associated with electron "orbits" the debate ranged on. There was a problem with many electrons competing for orbit levels. Then Pauli came up with the Exclusion Principle. This states that in atomic or subatomic situations, a particle can only occupy a free state and then somehow exclude all others from occupying that state. It was as simple as a busy seat theory. If someone is in a seat it can't be used by anyone else. This exclusion supposedly prevented electrons from sharing orbits. But there was one exception. Two electrons were allowed (if that is the right word) to occupy the same energy levels. To do this they each had to have one of two different "spins". This property called spin, which could have two states, was introduced to allow up to two electrons to occupy each energy level which was required to satisfy known electron shell counts.

It was then shown that all the orbit possibilities aligned well with existing atomic data in the periodic chart as well as with observations such as line spectra. These had already been shown to be related to electron orbit jumps and corresponding changes in energy levels. The exclusion principle solved more of the problems with the Bohr model and it went on to become a key component in further subatomic particle physics especially with quark theory. The full explanation of how spin and exclusion work is based on quantum wave mechanics or quantum theory and is beyond the scope of this work.

NEUTRONS—PAULI (1930)

The neutron was the last of the standard atomic components to be fully discovered. I say fully discovered, because Rutherford had already proposed such a neutral particle to make up for the mass of the atom for all atoms larger than Hydrogen. Perhaps this delay in the full discovery of the neutron was because it was heavy and easily stopped. It was also electrically neutral with no charge and could not be deflected by E/M fields. These properties meant that it could not readily be differentiated or observed in existing atomic scattering experiments. New methods of detection or capture were required to enable the neutron to be seen and for a better picture of the neutron to be developed.

Other concepts on how the parts of an atom behaved were also being proposed. At this stage the macro model of the atom was reasonably well developed and became the accepted standard model of the atom. This atomic model as shown in the above diagram was the one that was used as the basis for further work on atomic structure. It is the standard macro atomic model still commonly used today and is the one taught in schools at a basic physics level. It provides a sufficiently accurate picture of what an atom is for most low level science studies and it is the simple model used in the popular media.

Unfortunately, this elegant but simple model was soon found wanting. It couldn't explain many experimental results and observations. Many physicists then became actively engaged in the race to better understand the atom and improve the model. The problem was that the atom, and more so its components were so small and so difficult to work with that a new type of physics was required. More specialised technology was also needed to "open" the atom. Two key developments helped in this way, but before they are examined in any detail it is worth revisiting the atomic model to re-examine some aspects of its structure, behaviour and especially orbiting particles.

LINE SPECTRA

A number of famous physicists continued with research into the atom and especially electron activity and light. Some played a more significant role than others but I have only singled out a few who played fundamental roles in this field. Interest in this area of physics goes back even before Faraday. His bending of a candle flame by a magnetic field or was it an electric field, was typical of these early experiments. Many aspects of physics were being linked to electromagnetism and light. It is a fascinating subject and one which I feel sure has not yet been fully addressed. There are many good books on these topics if the reader wants to follow anything up. This is another area on which TOLG may shed some light in due course and not just CGBR or PP light.

Experiments carried out to measure the frequency of (electron) light from heated objects showed an amazing observation. The light was seen to be of particular colours. When this light was shone through a prism to examine its spectrum, discrete sharp bright lines of light were observed at different frequencies in the spectrum. These spectral lines were being emitted from the atoms of a heated object. Different elements showed up different spectral lines. In particular the gas hydrogen showed up distinct sequences of spectral lines. These were soon found to be related to multiples of the wavelength of an electron in the hydrogen atom. All other elements were soon shown to have their own spectral line "fingerprint". Spectroscopy soon added further experimental support to the study and development of the emerging theory of the atom and electron orbits.

Basically the theory, which is an extension of Einstein's photoelectric effect, shows that an electron in an atom or in a free state can be excited and then made to release energy. This energy is given off in the form of a photon of a particular frequency or wavelength depending on the amount of energy released. The amount of energy is related to the energy level or orbital position of the electron. As an electron jumps (or falls) down a level or transitions down through the various acceptable orbital or energy levels of an atom, a set of emission spectral lines is produced. Or in reverse, as the electron is energised and made to jump up through allowed energy levels of an atom a set of absorption lines is created in an otherwise continuos spectrum of light. Particular elements or atoms can then be identified by examining the spectral lines produced.

So what does this all mean to TOLG. Well in the chapter on the cosmos, spectroscopy plays a major role. At the atomic scale it is also important but I am not yet sure of any immediate relevance of TOLG to atomic spectral behaviour at least for electron light. As for proton light well that is a horse of a different colour or should I say much shorter wavelength. When light energy and spectra at proton level are eventually addressed, this will produce a whole new field of physics. The relevance of proton photons to TOLG is obvious. Obtaining line spectrum from proton activity will be an amazing discovery and not just because of the obviously strange and new technology involved but it will also prove that TOLG is the correct theory of gravity.

RADIO ACTIVITY

A lot of research was still associated with natural atomic activity or radioactivity as it was called. This led to the discovery of fission (heavy element decomposition) during the early to mid 20th Century. Fission is the breakdown of heavy elements into lighter elements with the release of energy. This discovery resulted in the development of controlled chain reactions or piles and also uncontrolled chain reactions for atomic weapons such as those used in the Second World War. Eventually atomic fusion (light element combination) was discovered and it was also first used in making weapons. Both of these new types of weapons created the subsequent east/west arms race which nearly resulted in Armageddon. The continued proliferation of these weapons and potential for misuse still threatens the safe and secure future of the human race and all life on earth.

The development of so called controlled atomic reactions was also applied for more peaceful applications such as the generation of power. Nuclear power plants, of which there are many in operation today, use fission to generate heat to generate steam to generate electrical power. It is about time a more direct and efficient method of capturing atomic energy was developed. This process is often called clean energy because it doesn't burn fossil fuel and doesn't produce CO_2. But it is not clean. Fission produces deadly atomic wastes with a very long lifetime so perhaps it may lead to a different type of global warming. These atomic machines can also have failures such as Chernobyl, Three Mile Island and some in Western Europe. These have shown just how dangerous fission can be. Have we controlled the atom or are we in danger of loosing control?

Atomic fusion, which is the combination of small atoms in heavier ones such as isotopes of hydrogen into helium, is supposedly much safer. It doesn't produce (known) radio-active contaminants and it doesn't produce any greenhouse gases. The use of Fusion power for peaceful purposes, especially power generation, has been the focus of considerable money and effort by physicist for over 50 years. The creation of cheap fusion power was just around the corner when I was a lad. But unfortunately the corner hasn't yet been turned and I believe it will continue to be a dream for a long time to come. Fusion reactions today in TAURUS or TOMACS or whatever the latest machines are called, involve very high energies to start them. They also usually produce temperatures hotter than the suns surface in the fusion reaction plasma "core". It seems as though there is something fundamentally unstable about fusion reactions. They are not as self starting or as easy to control as fission. Recently some physicists "discovered" cold fusion. It did not require high temperatures and did not produce high temperatures. The fusion was reported to have been "observed" in cold mixes of special chemicals. It was not readily repeatable and it produced no real useful positive energy output if in fact it worked at all. It seems that it was more mirrors and smoke than steam and it seems to have disappeared from the science scene just as quickly as it arrived. Perhaps a method of controlling real fusion will be found based on a process similar to that which enabled fission to be "controlled". But this time perhaps it will be based on speeding up and not slowing down neutrons to enable fusion to be started and managed. I am not sure what TOLG will do in this area but let's hope it leads on to some useful applications. Who knows, perhaps a PP laser will be the key to controlling fusion!

As technology became bigger and more expensive, developments in the atomic model continued. During the next 50 or so years, higher energies helped with the "discovery" of many new subatomic particles. Perhaps the most amazing was the discovery or perhaps more appropriately the creation of anti particles. These were proposed as being required for energy conservation laws and to explain observations of atom smashing experiments. Unfortunately the lifetime of these so called anti particles is so short (near the Plank time?) that they can't easily be observed or measured. They are primarily theoretical and are "created" to correct for unexpected particle transitions and energy transfers as well as to explain some strange phenomenon which were apparently unexplainable without them. I will not address such particles or perhaps I should say anti-particles any further. I am not sure what impact they have on TOLG or vice versa but perhaps TOLG will again shed some light on them.

By the middle of last century, the "solid atom" was well and truly smashed along with the planetary atomic model. The "solid electron" was still solid but the "solid proton" and the "solid neutron" had some explaining to do. It seemed from atom smashing experiments and associated new mathematical modelling that these particles may not be "solid" or fundamental after all. I am not sure what this means for proton photons but I don't believe it negates them in any way. New high energy atom smashing produced some strange "trails" which indicated the existence of other as yet unknown particles. Every new physicist on the block had to discover a new particle and soon a new particle soup was created or should I say cooked up, to extend the metaphor. Most of these new particles had very short lifetimes and they were not as stable as their bigger parents such as the proton or the neutron. It was assumed most came from these larger particles as the electron was still considered indestructible and there were no other obvious sources. Some of the new particles showed strange behaviour in electromagnetic fields and under other conditions and investigations into what they really are continue today. I will briefly mention some of them to try and complete the picture but I have to tread carefully again so as not to get out of my depth. I originally thought that some of these strange particles such as the neutrino or the new bosons may have been candidates for CGBR, but now I am not si sure.

QUARKS—GELL-MANN (1964)

With newer, bigger and better (and more costly) research technologies, new directions in the understanding of the atomic model occurred. New particles were being discovered or proposed including an array of anti particles and mystery particles that could not be easily explained by the current model. It was eventually discovered or more correctly proposed that the components of the nucleus of an atom, the proton and the neutron, may be further divisible, at least in theory if not in practice. The components were given the name quarks. One physicist is often called the farther of the quark model and that is Murray Gell-Mann. Different types of quarks were proposed and many were then "found" but they were eventually grouped into three basic types. Each of these types were said to have, or more correctly were assigned properties to differentiate them from each other. Strange names (literally) and properties such as spin, charge and colour or chrominance were created. Combination rules were also developed to enable parent particles to be constructed that matched observations and mathematical modelling. This work was similar in a way to the creation of the period table over one hundred years earlier. Fewer parts are involved this time or at least that is the current position but the number of particles is growing at an alarming rate.

My interpretation of the current evolution of these subatomic particles is shown below. This may not be an exact or up to date model but it should suffice for this discussion on the atom and on TOLG gravity and its atomic effects. It can be seen that as technology improved and the understanding of the physics and mathematics of the atom progressed, the atom has been slowly taken apart. Will it be like a watch and eventually become a set of basic components like a springs and gears or will the breakdown go on forever or at least down to Plank dimensions?

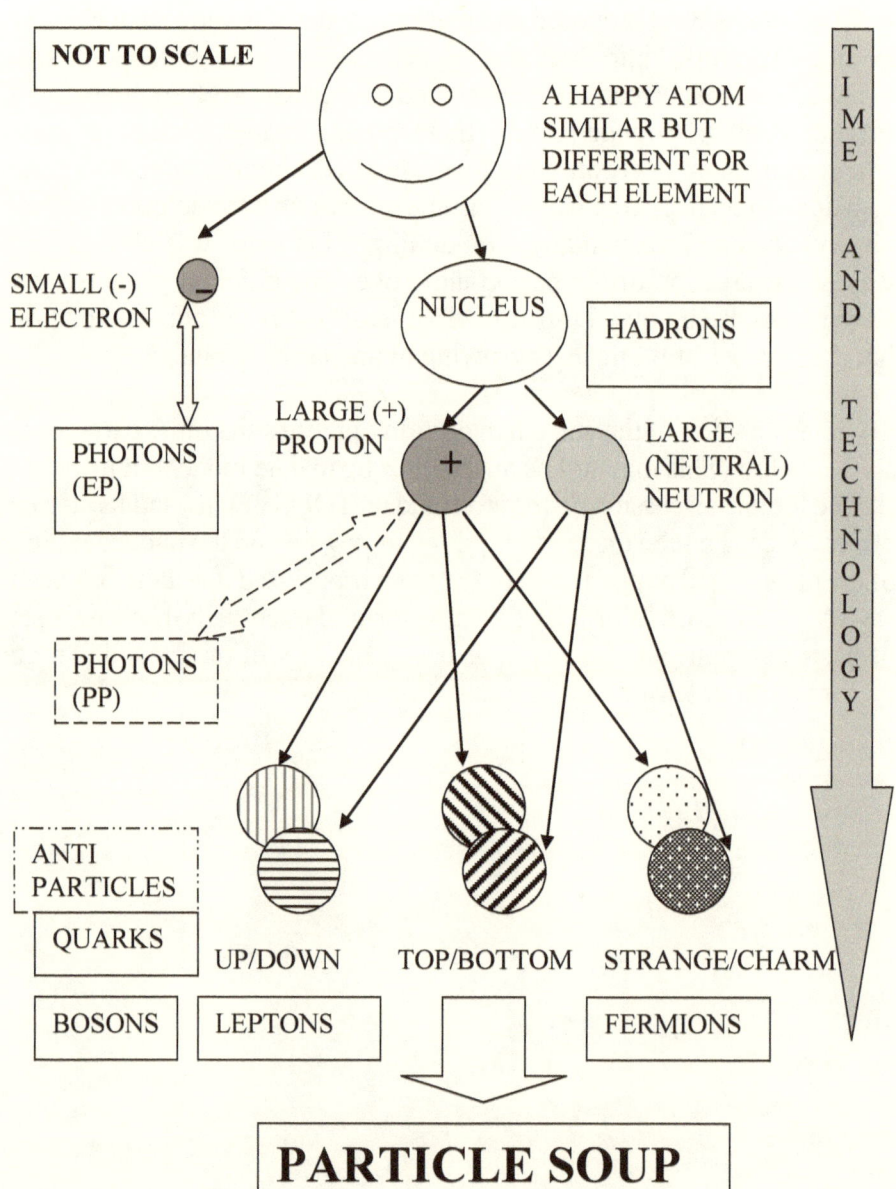

HADRONS, LEPTONS, BOSONS AND OTHERS

There are a large number of particles in the subatomic particle soup. These particles have been categorised into groups based on general properties such as size, mass, behaviour, lifetime, speed, etc. The main groups are; the Fermions which include the bigger more energetic and slow moving particles called Hadrons and their components; Leptons which are smaller, less energetic "slow" moving particles such as the electron; and last if not least the very small and mostly light speed particles called Bosons. There are many more including a vast array of anti-particles and ghost particles. Sorry if I have lost some readers but I am lost too in this increasingly complex quantum model of the simple old atom. And I have added my new path to create proton photons just like electron photons were originally linked to the electron.

So let's look at my new version of the old model of the atom and a new idea with protons that doesn't involve their destruction. Well not completely.

ORBITING PROTONS—PARKES (2005)

One feature of the planetary atomic model that doesn't seem to have been fully recognised, at least as far as my research into the literature goes, is that an electron can't simply orbit a fixed nucleus. The picture must surely be more complicated that that. The nucleus or at least the protons cannot be fixed. For example, let me use a cosmic situation to try and explain what I mean. Many people believe the moon orbits the earth. Well that is correct in a sense but in another way it is not. In reality both the moon and the earth orbit each other. In a way it could be said that the earth is also in a geostationary orbit of the moon because we always see the same moon face. To put it more correctly they both orbit about a common centre of gravity. This common centre of gravity is determined by the masses and the separation distance of the bodies. For the moon and earth this is well outside the earth. Therefore the earth also orbits this centre of gravity and hence the earth "wobbles". The wobble and is quite slow at about 28 days periodicity and I am not sure how observable it is by cosmic observations but I must follow this question up. I also wonder what this monthly earth wobbling means for the accuracy of some cosmic distance measurements using RADAR? I am sure it is taken into account.

This wobbling can perhaps be more easily explained by using another example. Let's say you hold your *young* son by the hands, assuming you have one. As you swing him out and gently swing him around, you will both spin around an imaginary point (centre of gravity—CG). This will be somewhere near your hands not your head. You will not simply turn around in a vertical way about a fixed point on the ground where you are standing but will need to lean out backwards and swing around. You will then rotate around a common point which is away from you and toward your son. So in a similar way, the earth and the moon rotate around each other so to speak. If the earth and the moon were of equal mass then the central point of rotation would be the mid point between the two bodies. This is all simple Newtonian mechanics, nothing new under the sun or the moon. It is a natural consequence of Newton's gravity and laws of motion and of course can also be shown to be an outcome of TOLG. For each action, e.g. the force of gravity from the earth to the moon which together with the moons speed makes the moon orbit and not crash into earth, there is an equal and opposite reaction, e.g. a force of gravity on the earth from the moon which makes the earth orbit the moon or at least the mutual centre of gravity (C.G.).

This concept, of course, also applies to the sun and the earth and to the sun and all bodies in our solar system. It also applies to all other gravity situations. The common centre of gravity for the solar system is not simply always the centre of the sun. While it is mostly near the centre of the sun due to its phenomenal mass, it changes a bit very slowly from time to time. It may even be outside the sun on occasions, especially when all the planets align on the same side of the sun. Using estimates of solar system masses and distances results in about 2 solar diameters for the maximum distance the CG can ever be away from the centre of the sun. So the solar system may orbit around a point that is outside the sun on certain extreme occasions. One day I must find out when this will next happen and do the full derivation for the worst case. This is another interesting solar system wobble for further study. Perhaps it has solar system stability implications as suggested by Newton and Poincare as well as potential earth quake causing implications for us here on earth due to the extra earth stresses that may be involved!

So what does this all mean for the atom? Well, if conventional physics applies (and I recognise this is a major assumption) then the electron and its "related" proton must both orbit a common point. Now I use the term "related" as they are generally in pairs in an atom but I am not sure what it means in current quantum speak. So, to put it in simple mechanical terms, the proton must also be in some type of orbit. It can't be fixed in the nucleus as per the original (and current?) model. Of course the orbital arrangements must take account of the relative masses or energies of each particle. And the orbits must also be compatible with "atomic orbital theory" based on wavelengths. In this theory all particles have a wavelength (wave particle duality) based on their energy content or mass and this wavelength determines their "orbital" behaviour. So each proton must also be in orbit, but with a much smaller orbital radius compared to electrons. The electron and the proton or each electron/proton pair must be in orbit about some common CG point to stretch the analogy a bit. Where this point or these points are, is perhaps not significant but it is still a good question. They may all be at slightly different places with respect to the so called centre of the atom but they must be near the nucleus of neutrons.

This can perhaps be best considered in a simple Hydrogen atom which consists of one electron and one proton but no neutrons. The electron can't just simply be orbiting a fixed proton. The two particles, the electron and the proton must both be orbiting each other or more correctly orbiting about a common (electrical) CG. I know there is charged particle attraction physics and other aspects involved but this doesn't change the overall situation, it should just confirm it. Therefore it seems obvious that both charged particles in the atom, the electron and the proton, are in orbit, most likely around the nucleus of neutrons or in the case of hydrogen, around some virtual mutual CG. Now I understand that the concept of an orbiting electron in a classical planetary sense is only that, a concept. But that concept was used successfully to try and explain the model in the early days. Electrons, and therefore also protons, don't just simply orbit a fixed nucleus like planets around the sun, there is more to it than that. The current theory of quantum mechanics proposes a different concept which involves probabilistic mathematics and other quantum laws of particle behaviour. Quantum mechanics, or is it QCD, proposes that electrons exist in clouds of probability. There is only a defined probability (less than one) that an electron at a specified energy level is in any place at any time. But there is no reason why this general concept of floating electrons can not also apply to protons. In fact it seems obvious to me that it must also apply to protons. This equivalence in atomic behaviour between electrons and protons must surely be the only explanation. So all atomic protons must also be "orbiting" or moving (shades of Galileo's comment—"and yet it moves"). How profound!

This type of balanced movement must apply to all electron—proton pairs in all atoms. The basic concept of movement or orbiting protons now seems obvious and can then be extended to any other type of atom even with neutrons. It would seem that each electron/proton pair in any neutral (non-ionised) atom must be orbiting each other. Of course this needs to be interpreted in terms of the original de Broglie model or Schrödinger wave theory or more correctly the current QCD probabilistic model of the atom and its components. But the basic concept is sound and it must have wider implications. The big question then becomes what does this mean to the current atomic model if anything. In particular what does it mean to our understanding of atomic forces and proton activity, especially quarks, proton energy states and proton energy levels and transitions? Does it mean that whatever the electron can do, the proton can do and perhaps do it better. Does it therefore automatically prove the existence of proton light or proton photons! It seems to me to add strong support to my proposal for light from protons or PP as I call them! Perhaps not complete proof but that is coming.

This concept of an atomic model with orbiting protons may have implications on the original Rutherford scattering experiment that "proved" the nucleus had a hard centre, just like some chocolates I have eaten recently, but perhaps not. His experiment was very crude in terms of estimating the size of the nucleus in an atom. It had no real way of determining any fine structure of the nucleus or of determining if protons were really fixed in the centre or not. That was all supposition. As orbiting protons have a very small radius of orbit due to their very high energy and short wavelength, they would still seem like a hard core to many Rutherford projectiles. I have tried to show how this may work in my new version of the old atomic model below. I hope am not perpetuating a simple and false model in any way but it gets the point across.

THE NEW "PLANETARY" ATOMIC MODEL
(AGAIN NOT TO SCALE)

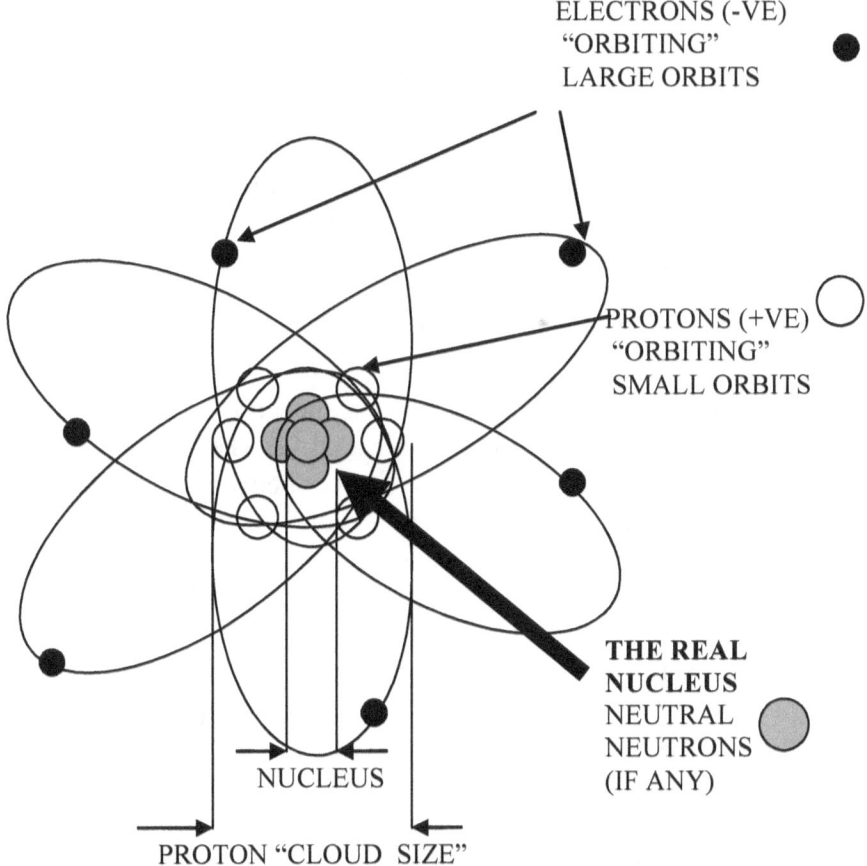

ELECTRONS (-VE)
"ORBITING"
LARGE ORBITS

PROTONS (+VE)
"ORBITING"
SMALL ORBITS

**THE REAL
NUCLEUS**
NEUTRAL
NEUTRONS
(IF ANY)

NUCLEUS

PROTON "CLOUD SIZE"

← ATOM (ELECTRON CLOUD) "SIZE" →

I have been unable to obtain any information on my idea of orbiting protons in any current reference material and will continue research into this area. Perhaps this idea has already been proposed and there are extensive theories on it and what it implies for proton activity. However, it seems that current quark theory and the QCD model suggest it hasn't been raised before and they challenge any such proposal. Research on this subject is most likely very leading edge and hence any documentation may be very difficult to access. Also any associated mathematics or theoretical modelling about this will perhaps be way beyond my mathematical capabilities at this stage. But it seems to me that such an idea fits in well with my theory that electrons and protons have similar properties and therefore share similar processes such as "orbiting" the nucleus or atomic CG and in the creation of light. Therefore, I now firmly believe my hypothesis that protons can and do create light or CGBR is correct and hence my gravity idea for TOLG is soundly based. It just needs to be proven. In the meantime back to the atom.

TOLG AND THE (NEW) ATOM

What are the implications of TOLG on the atomic model or more importantly what does the atom mean to TOLG? The model has a large number of components, various types of forces between them and a range of interactions. On a larger scale, gravity is one of these but gravity may also work on a smaller sub atomic scale. Other forces have been proposed to control subatomic activity and gravity is not usually considered to be active at this level. But TOLG gravity may fit in here, I believe. The four established forces in physics are gravity, electro-magnetism, and the strong and weak nuclear forces. Gravity is or will be explained by TOLG. Electro-magnetism includes charged particle actions as well as magnetic actions if they really are different. Gravity and electro-magnetism are related in many ways. The strength of the force is reciprocal and is directly proportional to mass for gravity and electromagnetic fields for E/M forces. They also follow the inverse square law and both act at the speed of light. The forces also apply over infinite distance. But is there any limit in terms of very short distances? Of course TOLG relates them directly as both are caused by light. Gravity through TOLG is due to photon pressure (CGBR) while E/M forces are due to some unexplained field effect or perhaps photon interaction, possibly also involving a pressure effect, who knows?

The other two forces, the strong and weak nuclear forces supposedly only apply within subatomic distances or have an ultra short distance of effect. I have been unable to determine what other rules apply or if there is a distance law such as an inverse square law but will continue my research. It seems that both of these forces don't apply at distances greater than atomic size but it is not yet clear why this is so, at least to me. It is also unclear to me if any other rules apply such as speed of action or any quantised magnitude relationship. The question as to whether the forces of gravity and E/M fields apply to subatomic particles at short distances is an interesting one. The electron and proton are oppositely charged electric particles and should follow the simple rules of E/M force or Coulombs law. All these particles also have mass so some form of gravity may apply between them. But I am not sure how G.R. gravity applies within subatomic distances, if it applies at all. I have been unable to find any simple explanation of how G.R. works inside the atom but a more important question is how TOLG works at this level.

As TOLG is based on proton light this would suggest that TOLG has a role to play within the atom and at what are normally called subatomic distances. It would seem from first principles that TOLG would impact on protons and perhaps neutrons, just as electron light does with electrons. At atomic scales EP create simple pressure effects through elastic collisions at atomic size or electron cloud level. There are perhaps two types of photon-matter interaction, resonant and non resonant. Resonant actions are perhaps similar to the photoelectric effect. This is when a photon has the appropriate energy (wavelength) and is captured by an electron (harmonic capture) which is then jumped up one orbit level. For photons that are not the correct wavelength for the atoms involved, perhaps an elastic collision occurs and only a pressure effect is created. Atomic level activity from gravity photons (PP) may be such that energy is not (easily) transferred to atoms and they just have elastic collisions. Therefore, matter doesn't heat up from the impact of gravity photons; they only impart momentum, not energy!! So CGBR could be said to be cold, but why? Why does it bounce off in "perfect" elastic collisions and only impart momentum but not transfer energy to matter (atoms) that it strikes. Of course it may impart some energy and this may be associated with unexplained internal heating of some bodies (e.g. the earth) but it would seem that most CGBR is reflected.

So the frequency and hence energy level of a gravity photon is so high that it can't be absorbed at atomic electron level. It can't "see" electron clouds and just passes through each atom until it bounces of at nucleus level where it imparts momentum. That is also the reason why most gravity photons pass through most material with no impact (no deflection/reflection) and there is a high degree of penetration. Of course this is a simple view and needs much more in depth analysis of the quantities, energies and wavelengths involved. In particular if gravity photons are created by protons the question must be asked, how are they absorbed again? This is another mystery and is probably related to the atomic binding process (strong/weak forces) which makes it difficult to open up or smash an atom or the nucleus. To break open Hadrons requires extremely high energy. This is a "weak" area of the TOLG gravity theory at present but further work will hopefully solve it. Again a call for super CERN.

SUMMARY

The likely impact of TOLG on atomic physics seems to be less significant that the likely impact of new atomic physics on TOLG. The outstanding problem with TOLG is finding, isolating, detecting and measuring the light that causes it. This light from proton activity, called proton photons or PP or CGBR will be hard to "see". These photons have very high energy and very short wavelength. Detection by existing electron based (EP) technologies appears to be problematic. This is because it seems that electrons will not be able to see PP light, they are the wrong "size" or they have the wrong capture area. A new approach to PP light is required, perhaps based on using unstable radioactive atoms as detectors. It might be a case of spontaneous emission from radioactivity not really spontaneous but in fact is being triggered by random strikes from PP. Anyway let's move on and keep these ideas in mind. Let the search begin for the real story of the atoms role in PP and vice versa. Perhaps my new atom will come to life.

CHAPTER 8

Cosmic Implications

One way of looking at a new gravity theory such as TOLG, is to consider the implications it may have on certain areas of physics where gravity plays a key role. This chapter looks at the importance of TOLG for the macro or large scale end of physics, called cosmology. There is no doubt that in this game, gravity is king. So TOLG obviously could have a big impact. And while the man focus is on the bigger picture of the total universe, I have also included a section on our own solar system as a special case. Even though we supposedly know a lot more about our own local cosmic environment, there are still many unanswered questions about it. Existing gravity theories don't seem to be able to answer all of them. Does TOLG provide better answers? At this stage this is open for discussion but perhaps it does in some ways. When TOLG is completed and accepted as a sound theory, work can commence on trying to answer such outstanding questions using this new a comprehensive theory of gravity.

Back to the bigger picture, cosmology is currently the focus of considerable research effort. The best scientific understanding or should I say the current "official" scientific position on cosmology and the universe is well documented. Scientists working in this field are generally more than willing to air their views at least in general terms. This is also a very popular and controversial subject, so these views are usually sought out and well presented in the media. However, almost every time a new idea or theory about the cosmos is proposed it seems to raise just as many questions as it answers. A rush of counter ideas and challenges or at least alternative theories follows each new announcement. While these differences of opinion are all the better for science, this myriad of interpretations seems to further confuse the picture. It seems to indicate considerable uncertainty in some of the fundamental aspects of our understanding of the universe. The cloudy picture of the cosmos just seems to get cloudier by the day.

Of course, the key areas of cosmology that are significantly impacted are those that are based on gravitational effects and especially interpretations of General Relativity. As TOLG is based on light, those aspects of G.R. which make proposals based on the impact of gravity on light may need to be carefully reconsidered. This is because a major outcome of TOLG is that gravity is caused by light and therefore it should have no effect on other types of light. This also has serious implications for the Big Bang theory and the supposed origin of the universe as well as Black Holes. Other aspects of cosmology that are based on the behaviour of light also come into question because of this new theory of gravity. These areas are reviewed and possible implications are suggested. In more general terms though, there may be a need for new interpretations of some experimental results and research when viewed from the point of view of TOLG.

MODELS OF COSMOLOGY

Any work on cosmology is usually conducted within a framework that describes the current scientific theory of the universe. This current theory is often contained in something called a standard model. This becomes The Standard Model when it is widely accepted. I have researched various aspects of cosmology and I have examined the current or at least a reasonably recent version of the current Standard Model of Cosmology and tried to develop a good understanding of it. I wanted to determine how important gravity was and in particular what impacts a significant change in gravity theory may have on the model. While I am not an expert in this area ("where have I heard that before", you may say) I found gravity was a critical component of almost all aspects of cosmology, as would be expected. Therefore any new gravity theory such as TOLG would have serious implications on the model. It won't change the universe or how it works but it will significantly change how we interpret and model it. Potential impacts of TOLG are everywhere.

The current model is of course based heavily on General Relativity and in particular the Big Bang theory. The work I have done on gravity and light immediately indicated a conflict with this idea and made accepting this part of the model problematic but let's continue. While the standard model is quite well developed, it is also constantly changing. New ideas, observations or calculations regularly surface that challenge some broad aspect or component of it. So, while it is supposedly complete there are many parts of it that raise more questions than they supposedly solve. The model I will consider here is based on one that seems to be reasonably stable and self consistent. It may have changed since I started work in this area but hopefully not significantly. I have tried to identify those areas of conflict with TOLG or that seem to be inconsistent with my current view of gravity and of cosmology. At this stage I am sure many cosmologists have already "pulled their hair out", that is if they continued to read the book up to this point. Words of disbelief or even stronger words are probably already being muttered at my bold statements and challenges. I can already hear them asking questions such as "who I am to challenge such a massive body of scientific wisdom".

But before I answer that question or look closely at the current standard model of the universe it may be useful to look at some of the history behind it. There has been an ongoing evolution of ideas on what the universe is and how it came into being for many millennia. Questions about the future have also been raised such as what is going to happen in time to our planet, our sun and the complete cosmos? Will they ever end? Many of these questions have been answered by saying at least nothing will change in my lifetime or adopting a religious approach to nature and the universe with an infinite creator making and controlling everything. Unfortunately these types of approaches often raise further philosophical questions about who created the creator and so on and it becomes a never ending circular story. A bit like the Big Bang itself.

Models of cosmology have come and gone. Some stay for a longer period than others and some are more complex than others. But they all have problems, none are perfect. Some were based on primitive ideas and some have been based on complex mathematics. So let's review the major ones and then inject a new one just for discussion.

THE PTOLEMY MODEL

Perhaps the most well known original model of the universe was the Ptolemy model. This model was quite simple and it was relatively static and unchanging in trems of acceptance for a relatively long period. It had the earth at the centre of the universe with all other "heavenly" bodies rotating around it. It had to fit the philosophy of the time as well as the daily or should I say nightly observations of the cosmos. The model tried to explain our universe based on the simple human understanding of how things worked and to a large extent on the egotistical human belief that man is a divine creation. The model was believable, logical and rational in it's day. The key ingredients were;

1. Humans and earth play a unique role in the universe,
2. The earth is very big and doesn't (seem to) move,
3. So the earth is at the centre of everything,
4. The planets (night skies) rotate (seem to) around us.
5. Everything was fixed to concentric crystal spheres
6. Epicycles were created for unusual movements.
7. It included the universe.

This model was also based on some existing myths of the time such as a concept of perfect shapes such as spheres. This was used to support the idea that heavenly crystal spheres encircled the earth to support the cosmos. These crystal spheres had the stars or planets attached to them which were also perfect spheres. New observations of planetary movements challenged this simple circular orbit model. Special movements called epicycles were then added to try and explain them. When even these strange and complicated additions could not explain all observations, other more complicated movements based sub spheres were added. The model became complicated with new bits being added as observations showed up inconsistencies. Sounds familiar doesn't it.

This ad-hoc "patching up" of the model seemed to work for almost 1500 years. But this initially sound but by now somewhat patchwork model which was based on scientifically unsound foundations, eventually became unacceptable. It was challenged by the questioning minds of the new age of science that was emerging during the renaissance. The complexity of the model with its epicycles and sub spheres eventually became too complicated to modify any further. It simply became unusable. Unfortunately any questioning of the model was difficult because it was part of state and church sanctioned doctrine. Challenging it often had serious consequences literally for life and limb. Perhaps this was part of its survival mechanism.

There was a growing need to better answer the increasing number of anomalies and to bring about a rational new approach to modelling the cosmos. Eventually there was a fundamental shift in understanding of how the universe or at least the solar system worked. This led to the collapse of the Ptolemy cosmic model. But it required perseverance by strong minded people to completely destroy it. This change was revolutionary and it also challenged the state and church hold on society and in particular on science. More background to this challenge is given in the history of gravity.

Perhaps the most important single player in making this jump was Copernicus. He really started the ball rolling toward a new model of planetary movement and the solar system. Another key player was of course Newton. He developed a comprehensive mathematical model for the solar system and provided a new scientific approach to modelling the cosmos. His universal gravitational force between any bodies could easily explain orbital mechanics. It became the mathematical basis for a completely new cosmic model and also helped find new planets and other bodies.

THE COPERNICAN/NEWTONIAN MODEL

So a new model of the universe was eventually developed and accepted. It was based on a number of new ideas, theories and observations. The key ones were;

1- The earth, sun (and humans) are not special (Copernicus),
2- The massive sun is at the centre of our solar system (various),
3- The earth (and all planets) really move (Galileo and others)
4- Planets move in elliptical orbits with an equal area law (Kepler)
5- The planets (and all cosmic bodies) obey gravity (Newton).
6- The model applies to the universe but little data available.

These factors are roughly in order of occurrence but not necessarily in order of importance. While the development and acceptance of the new model was well underway by the time of Newton, his work was perhaps the most scientifically significant. It ensured the new model had a sound mathematical basis and this enabled new observations and discoveries to be made that further developed the cosmos.

This new model of the universe was seen as a paradigm shift from the old model. It worked well and had a sound scientific foundation. Calculations of masses and distances showed how big the sun really was and how far away it was and how much further away some of the planets were. It shed new light on the vastness of the cosmos and of the possible types, sizes and distances of other cosmic bodies. The solar system as we now know it was born or at least a better understanding of it was made in terms of known physics. And the universe suddenly expanded or at least our understanding of it did.

It is interesting to note at this point that Newton supposedly had a problem with this model in terms of long term stability. He seemed to believe that the solar system and the universe could become unstable under the application of his universal law of gravity. One of the possible outcomes was that the universe would eventually collapse. I don't know much more about this but it is interesting to note that apparently Newton had thought about the wider implications of his theory of gravity on the universe. Perhaps he realised there was a need for a more comprehensive model of how the complete cosmos worked. The famous mathematician Poincare, better known for his folding manifold conjecture proposition which was only recently solved, also worked on this gravity stability problem and supposedly came to a similar instability conclusion. It seems that not much more thought has been given to this issue and I am not sure of the current view. It has most likely been swept up in G.R. and is part of the latest cosmic model.

The universe or at least our understanding of it then seemed to settle down again for almost three hundred years. It became accepted dogma, just like with the Ptolemy model before it, but at least is was more scientific and it raised fewer problems. A new era of complacency and satisfaction took over. There were some unanswered questions due to minor deviations from predicted orbits of some of the planets in our solar system. However most of these were settled when missing planets were postulated and finally found. Other questions about the bigger picture and about the extent of the universe and its stability only just bubbled away in the background with low interest. Questions about the size and age of the universe were just curiosities. At this stage it was generally believed that the universe was stable and infinite. It seemed that there could be no more surprises and that no big changes to current thinking were required. In any case, physics, mathematics and technology were not yet developed enough to enable new problems to be identified, raised or addressed in any significant way.

Eventually even this very sound and scientific cosmic model came under more and more scrutiny. This challenge was made somewhat easier by the fact that the model was no longer part of fanatical state or church doctrine. Criticising the existing cosmic view did not automatically lead to disastrous consequences. However the model was based on the theories of some of the most famous physicists that had ever lived, so to challenge it without very strong proof may hot have been good for one's career. While the existing model was widely supported by the scientific community, physics had come of age as a more open forum. Scientific challenge was part of the norm. As holes started to appear in Newtonian gravity theory, especially with its infinite speed hypothesis, the cosmic model was also found wanting and a new approach was required. Again observations and mathematics had caught up, based on the work of a new generation of physicists.

Then along came Einstein with his theories of Special Relativity and a bit later, General Relativity. Also about this time our understanding or perhaps more correctly our lack of understanding of the universe and its vastness, was increasing. A number of factors had a significant impact on the study of cosmology. Cosmic distances, times or ages, energy density and structure, were being questioned and examined as never before. These developments changed our understanding of the universe yet again. Perhaps the biggest factors were the implications of inertial relativity and the absolute speed of light as well as the new theory of General Relativity and the proposed space/time warping based on mass/energy densities. This new gravity theory had to be factored into our understanding of the universe somehow. A new picture or model was required. Here we go again.

This mathematical model of gravity developed by Einstein, relates spatial mass/energy density and space/time curvature to movement of a body due to gravity. It was without doubt a phenomenal insight into a way to view reality. Shortly after Einstein published his theory, the mathematical model was worked on and "solved" by some of the great names in physics. A number of physicists came up with several key outcomes. One of the most famous was an expanding universe. Einstein didn't accept this outcome because he originally believed the universe was static. He therefore changed his G.R. equations and inserted a new factor (lambda—Λ) which effectively added extra energy (or is it matter or dark energy or dark anti-matter?) to make the universe static. Then everyone thought all was well again but a new challenge emerged.

One factor which challenged the old model and was used as the basis of the current cosmological model is red shift of distant cosmic light. A red shift was noticed in light spectra from distant bodies. This was initially discovered by Hubble about the time he discovered variable cosmic light sources could be used to estimate cosmic distances. These led to Hubble's law relating distance to variable light behaviour. This distance relationship was also connected to the observed red shift which was then proposed as being a Doppler shift due to the movement of bodies relative to the earth. Almost immediately red shift was interpreted as being caused by an expanding universe. At first this explosive idea was very controversial. But eventually it became accepted, especially by Einstein who visited Hubble at an observatory to see it for himself. It went on to become one of the cornerstones of the current cosmic model. The new observational data and expanding universe proposal had immediate impacts. Einstein changed his mind again and supported the idea that the universe was not static. He then proposed that his new fudge factor be removed to match the new data, saying that to insert it in the first place was his biggest mistake. Perhaps his biggest mistake was that he didn't really understand what Λ meant or what value it really should have.

What did these new ideas and factors all mean to the existing model of the universe? Obviously they had significant implications for the old Newtonian gravity based model. A new model was required, based primarily on the new mathematics of G.R as this was now the accepted scientific solution to all gravity problems. However, even as Einstein himself said, this new mathematical model of gravity based on G.R. was extremely difficult to solve but some had done it.

Once again evidence against the existing model was building. It needed be changed but what was the real scientific basis for the change. The main factor was the new theory for gravity called G.R which related energy, mass, space and time. There was also the interpretation of red shift as an expanding universe. So a new model was proposed which was quickly supported. It simplifies to the old Newtonian model in special situations such as for our solar system (or most of it) but adds new dimensions to the understanding of the cosmos. It is now accepted as the best fit cosmic model. Or is it?

THE (CURRENT) "STANDARD" MODEL

The current standard model of cosmology is again based on several key aspects of physics. These include a mix of solutions to G.R. theory and interpretation, a vast array of cosmic observations with their interpretations and also other supporting mathematical analysis. The main factors behind the current model are;

1. General Relativity (space/time warping) for gravity (Einstein),
2. Red shift (supposedly) due to expanding universe (Hubble),
3. Expanding universe (epochs of expansion, wound backwards),
4. A Big Bang of "infinite" energy in "infinitely" small space,
5. A finite, flat, homogeneous, isotropic universe (closed or open?),
6. Background radiation (CMBR) as Big Bang "echo",
7. "Dark" substances to solve problems (epicycles?).

The overall model, especially the mathematics is beyond my analytical ability and perhaps my comprehension at present. I would need to spend almost a lifetime of study to understand it and I am not sure if I have the time anymore. It is also far too complicated for a general discussion book like this. However, some points are well worth consideration so I will address those aspects of the model and try to explain some possible outcomes. Note that I use both the words current and standard to describe the model. While both of these are similar in meaning, they are both used in this context to make the point that really nothing is as it seems.

Everything is still uncertain and is constantly under challenge. It is just a theory as they say.

The main component of the model is G.R. and while this theory is almost universally accepted, I do believe there is room to challenge aspects of it because of some of its "predictions". One of the most important outcomes of G.R is the proposed bending of light due to gravity. This outcome was in fact used by Einstein and others to initially "prove" the theory was correct. Another is the so called problem of discontinuities that G.R. can't handle especially at very small scales. These arise at the beginning of the universe in the initial stages of "creation" and of course there are Black Holes. There may be other areas where G.R. doesn't seem to provide complete answers but I am no expert.

One discovery that helped form the new model was red shift. The view was that red shift was due movement of the source away from the observer based simply on so called Doppler Shift. This shows a mathematical relationship between the source frequency and measured frequency for longitudinal waves, for a body moving relative to the source or the media. Now as I explained previously, there can be no longitudinal based Doppler shift for light as there is no speed change, no media and light is not longitudinal. However red shift is still almost universally accepted as a direct indication of movement or increased separation between source and observer, based on simple longitudinal Doppler theory. I must make it clear that I have no difficulty with the fact that shi(f)t happens. There is also no doubt that red shift seems to occur right across the cosmos and is somehow related to distance. It has been extensively plotted by large scale observations of the cosmos by some of the best observatories. Interestingly there have also been a few observations of blue shift which complicate the picture. I also accept that there seems to be a strong correlation between red shift and light transit time. Now of course it must have a scientific basis but to directly relate it to movement via a Doppler effect may be incorrect and it needs more careful consideration. More on this in my next book on light.

The current standard model explains red shift in terms of expansion of the universe. The rate of expansion of the universe space/time is based on distance. The current red shift model describes how such cosmic expansion, which is dependant on the distance between source and detector and therefore also on the age of the photons, expands and therefore, reduces the energy of the photons. This expansion of photons is then directly related to the expansion rates and hence velocities and distances in a similar way to Hubble's Doppler proposal. Of course this expansion proposal also raises many more questions such as why don't other particles such as atoms or other simple atomic components of atoms such as electrons or quarks also expand or age in the same way and also loose energy. If all particles and all energies also change in the same way would red shift be detectable? Perhaps not, but is it so obvious they don't also expand? What are the relativistic implications here? Yes, out of my depth again, but perhaps I am on to something.

The next factor in creating the current model used red shift and the expanding universe idea. The expansion was wound backwards, or reversed, assuming that such things are reversible. Now it is well known that time can't simply be reversed but let's stay with this idea for a while. The conclusion was that all the cosmos was once closer together and even all at one place at one time in the distant past. This was used to imply that the entire universe was created from a single event which perhaps happened at a single point in space and time. Somehow this point of nothing with effectively infinite density suddenly all exploded outward at this time for no apparent reason and it continues this "outward" expansion even today. This Big Bang proposal was supposedly originally coined in a derogatory way by critics of the theory such as Hoyle. But the "in your face" name stuck and is now much revered by the cosmologists who support this theory.

So it was assumed that some long but finite time ago all the matter and energy and everything in the universe was together at one place (which didn't exist) and at zero time which also didn't exist. That is the fundamental conceptual or is it the philosophical basis of the current model. This single point of space and of time (that didn't yet exist) then became the instant creation of everything. This aspect of the current model fascinates me. The concept that it all started from nothing at all which was everywhere and that there wasn't even anyone or anything to know there was nothing, is real science fiction to me. It sounds strangely familiar to that other recent mad comment about war that "there are things that we don't know that we don't know"! Suddenly for no apparent reason there was an event or explosion which created an expanding "ball of fire and matter" that came from nothing and nowhere and suddenly became everything and everywhere. The universe started as nothing, or was it only almost nothing, and this led to the creation of everything we can see today and of course all that we can't see. Yes a "proof of God" for the "creationist" but where was God before everything was created? And now again, what we can't see has become a real problem.

The model starts at a zero point or discontinuity or was it just an extremely small location somewhere (or was it everywhere or was it nowhere?). It assumes an initial explosion of some type from something called a singularity (or is that only for Black Holes?). This singularity contained all the matter (and anti matter) and energy (anti-energy?) and time? (anti-time?) for the universe. At this point there was infinite energy density but not quite because infinite means a discontinuity and they aren't allowed in G.R. and physics or are they? This all happened at the start of time or zero time whatever this means and is this another discontinuity or singularity? This "fireball" then expanded "outwards" in space and time. It also created space and time as it grew (or was it all created instantly?). So initially there was no space and time then suddenly there was and it is still being made as we speak or should I say as I write and you read, or is it?

According to this big bang model there is supposedly a finite age or lifetime of the universe. Soon after the new model was proposed early last century, the age of the universe was initially calculated from the best observations of the time. However, it was found to be younger than the estimated age of the earth. This raised serious questions about the model but it was still supported. It was proposed that some of the observations were wrong or that factors used in the model such as Hubble's constant were wrong and needed changing. As technology improved and measurements become more reliable, more factors were changed and the age was increased to a more acceptable figure. We (the earth) weren't older than our father (the universe) after all. This lifetime is regularly recalculated from the latest estimates of the most distant observed objects in the known universe and is based on the time taken for this light to travel to earth.

It is assumed that these times are representative of the time for light to travel outwards from the "centre of the universe" (perhaps not a good term as there is supposedly not a centre in the current model, but it makes the point) to the edge of the universe (again not a good term because there is also not a real edge to the universe). But is the time taken for light to cross the universe really a good indicator of lifetime? And if so how do we measure the real cross universe light travel time? The interesting question is why the age of the universe is simply the time taken for light to travel from the current observed edge of the known universe to earth? I am not sure why this simple relation was used and why it was not doubled (or is it?) or at least increased by a factor to allow for the obvious conclusion that we are not on the edge of the universe, or are we? Yes I know the model tries to explain this idea a bit more in terms of its openness or closed-ness or flatness or saddle shaped roundness of the universe, but it still raises questions in my mind. Perhaps I must try harder to see the emperor's new clothes.

As this model was further developed and matched with new theories (such as the latest version of Λ in G.R.) and the latest cosmic observations, new questions arose as to how some situations could have occurred. One of these is the so called in-homogeneity or imbalance in the density of initial energy and matter. This explosion and the subsequent ongoing expansion were originally though of as being completely uniform but this didn't work out. It was too perfect for the creation of galaxies and stars and yes even us on our earth, to happen. The Big Bang expansion was then proposed as possibly being non uniform and it was calculated it needed to go through various stages of expansion at different rates or at various speeds or accelerations and for various durations. These stages are called epochs (not epicycles) in the current model. This epoch based development needed to happen in burst (jittering) with high speed expansions interspersed with slower development phases supposedly to allow for the development of the correct amount and distribution of matter and energy and other things in the universe. Without jittering things that have already happened in the universe would not have happened. This imbalance is required to eventually create real bulky matter such as stars, planets, galaxies and other known cosmic bodies. Other questions relate to the speed of expansion of the universe and the transfer of information with light speed always being a challenging limitation. Sometimes even this limiting speed is exceeded! Or has that idea changed? Then there is the need for the model to somehow create all the particles and matter that we know in the proportions we observe today. Nuclear synthesis was developed to handle this but it seems this sub-model can be easily changed to match whatever ratios are required. And is this also a reversible process?

NEW COMPONENTS OF THE
CURRENT STANDARD MODEL

The current standard model, called the Lambda CDM model, has a quite formal structure in terms of parameters, components and ratios or measures. If the reader wants more detail about this model and its components there are many good books on the subject and of course there is the Internet. I will not go into much detail on these analytical aspects of the model but let me raise a few interesting general points (and eyebrows?) about it.

The major component of the model is well known matter. This is relatively easy to comprehend at least at first. It is what we stand on and what we are made of. It is the sun, the earth and all the known (and as yet unknown but similar) physical cosmic bodies in the universe. So far so good. This matter is based on the atomic substances we all know and love, neutrons, protons and electrons (with quarks and other newly named particles in there somehow). However, this is where the main problem starts. According to estimates this matter occupies only about 5% of the universe. But the estimated mass and the observed rotation speeds of galaxies is inconsistent with that expected by gravitational force or G.R. calculations. Other matter was required to keep the galaxies in place and orbiting at the correct speeds. The term dark matter was coined to meet this requirement. This dark matter had most of the properties of ordinary matter but it could not be detected. It was called dark in that it supposedly didn't give off any radiation to enable detection, and the name stuck. Perhaps it could have been called invisible matter but that sounds crazier. Can I hear "not more epicycles" or is it "new clothes for the emperor".

Then another observation indicated a greater expansion of the universe than previously allowed for in the model based on new large red shift observations. The universe seemed to be getting, bigger faster that previously thought. This rapid expansion created a need for a new force and a new form of energy was proposed called dark energy in a similar way to the dark matter that had already been created. This dark energy would create negative gravity to explain his additional expansion or was it just another change in the cosmological constant. Now I must apologise if I have got the model and the problems and the answers all wrong. As

I said I am no expert in the current model of the universe. I am merely trying to fit some things together as an amateur physicist. Yes I can hear the experts say "don't touch if you don't understand".

Mathematics is generally at the core of all (both) extremes of physics, especially cosmology. This is because it is not easy to carry out experiments or think about possible cosmic events and search for observations to prove a specific theory. Such observations are either very hard to detect because of their rareness or to interpret because of their complexity. Many factors may be at play in causing any one event. There are also many aspects of cosmological physics that can not easily be researched. They are either beyond current technology or involve expensive resources and equipment. Recent advances in space based research as well as earth based technology, such as the CERN accelerator are moving slowly in this direction. It has been widely proposed that CERN may help in areas of cosmic research such as dark energy or at least provide a better understanding of the current model. Perhaps they will just end up adding more questions or is it epicycles!

While I call myself an amateur physicist I fall back on my engineering background. I try and relate all my experiences and theories to a more meaningful or perhaps I should say up front, a more practical framework. Straight away I can hear the specialist G.R. physicist say;

> ". . . but we can't always adopt a practical approach in theoretical physics".

I can also hear them say;

> ". . . we can't simply rely on our basic perceptions of reality to develop a comprehensive and realistic picture of the cosmos. Nature may play tricks and deceive us when we do that". Where have I heard that before?

Then there is;

> ". . . any model must primarily be based on mathematics, otherwise it becomes far to loose to define and work with".

Well, yes I agree there is some validity to these arguments but theory based primarily on mathematics may be just as dangerous. Some of the assumptions used in creating a mathematical model and some of the fudge factors then used to make work, may jeopardise the model and change the picture considerably. This is happening to some extent in G.R. with Einstein's fudge factor, his cosmological constant (Λ). While I agree on the necessity of sound mathematics, I also strongly believe that a requirement of any good scientific theory is simplicity. Newton's model was simple, but perhaps it was too simple for anything other than the solar system for which it was developed. Now perhaps G.R. or at least some of its interpretations with or without a cosmological constant are too complex even for the complete universe.

The existing model seems to be getting too complex. Again, just like the Ptolemy model, it seems to be going through too many epicycle developments. Can it be extended and supported for ever? Is it based on uncertain foundations, especially G.R. and expansion? TOLG challenges these and others such as the bending of light by gravity. My work has also led me to challenge the (tenuous) relationship between observed or measured red shift and universe expansion. Is there really acceleration of objects away from earth or the earth away from other objects? If TOLG challenges this red shift interpretation it also challenges the basis of the current Big Bang model. I just don't believe the universe "happened" that way and believe TOLG supports this view. So here we go again. First we have an old model that needs more and more epicycles to stay in touch with some form of reality. This model is supported by a theory that may be found wanting and may need replacement. Then we have a new gravity theory upon which to start a new model. Sounds familiar doesn't it. Well why not! Perhaps the old (current) model and theory has had its run. Let's see; model 1 went for about 1500 years and was defeated by science; model 2 was much better but not perfect and went for about 250 years. It was eventually "beaten" by better physics. Then model 3 emerged which again worked for a while and added a lot to our knowledge. But while it has lasted for nearly 100 years it has also been found to be wanting. How long can it be propped up by "epicycles"? Will it be replaced by even better physics?

Well, a new model is emerging based on TOLG. It is not yet a fully fledged model and is still in draft form and while it lacks the mathematical rigour of some earlier models it has a sound foundation. Perhaps it is at the stage of the Copernican proposal in model 2 or the "red shift expansion theory wound backwards" stage of model 3. But it definitely deserves consideration, even at this early stage of development. So here it is in all its (non) glory. Please consider, as the saying goes.

NON STANDARD MODEL OF COSMOLOGY

Any new model is generally called a non standard model so I have included that in the name. My new model is also based on a set of "scientific" assumptions. These are simpler or more general in some respects that those used in the current model and that must be an advantage. They may not yet be fully proven or even developed in a comprehensive mathematical way but they are just as sound as those used for other models.

The key assumptions are;

1. The earth, solar system and galaxy are one of many,
2. The universe is static, isotropic, homogeneous, infinite and ageless,
3. The universe is not expanding and did not start (recently) with a bang,
4. Red shift occurs, but is (mostly) just distance related,
5. TOLG (CGBR) is the basis of all gravity forces,
6. CGBR (G) may vary (long term) over time, direction and location,
7. There is no "darkness", what you see is what you get.

The major difference between this non standard model and the current standard model is that it proposes the universes is not expanding and never has and is static at least on a very large scale. Also gravity, or at least its strength, could and most likely does vary slightly across the cosmos. These changes lead to a direct challenge to G.R. and also a direct challenge to red shift interpretations. Perhaps the other most general and significant difference is that in the standard model it seems that some parts of the universe, such as the distant edges (whatever that means) which are the newer parts (or are they the older parts?) are different from our local space environment. There are places that have extra matter or energy or something that makes them behave differently to our local environment. I am at a bit of a loss to understand this concept of changing properties depending on location just as I am to understand and explain the epoch based expansion. They both challenge my understanding of simple physics and seem to create a problem for the universality of the laws of physics, or is it just me?

Anther major point of difference is that in the standard model everything is rapidly moving away from some (central??) point of creation because it was all created in a Big Bang. The standard model also proposes that this movement is different in different parts of the cosmos and some bodies are moving faster than others Of course this raises an interesting question as to whether there is a place where this movement is zero (the edge and centre problem again). Yes I know, moving relative to what, but again this is part of the problem. I believe there is a formula in the CDM model for an expansion rate as a function of location but I am not sure how to apply this. This expansion presumably applies to our galaxy, the solar system, the sun and the earth (here we go again, everything moves). But even for our solar system this expansion movement must be very fast, perhaps a reasonable fraction of the speed of light! If this is so then it has serious implications for cosmic modelling even under TOLG But I don't believe it happens in this way. I don't believe the local cosmic environment is moving rapidly away from anywhere or too anywhere. We are relatively (absolutely?) static in general.

I am sure that interpretations of observations will further support this model once it gains some initial acceptance. I just hope new discoveries don't also create epicycles as they mean trouble for any model. But these are not the main problem with this new model. There is so much support and inertia with the current model that an upstart like me has little chance of succeeding. But I will continue my work on analysing the standard model and trying to find better answers to the anomalies that are constantly being discovered. The full comparison will require a mathematical model of TOLG. This model may be similar to G.R with some form of complex field equations to determine the mutual shadowing effect at the speed of light. But it will not have a heuristic gravity constant and it will not have undefined fudge factors to enable it to match the latest results. Hopefully it will have no unknown factors or epicycles and be based on parameters for light, matter and energy and of course CGBR

THE SOLAR SYSTEM

So now let's look closer to home. While much of the emphasis of new cosmic scientific research is on the bigger picture of the universe, it is still very interesting to look in our own backyard, our solar system. Newton's laws of gravity provided a great leap forward in our understanding of our solar system. The model of a sun centred or heliocentric planetary system based on Newton's mathematical law of gravity is still well accepted. The model is still able to explain almost all aspects of the known solar system with a high degree of accuracy. The orbits of all the known planets as well as known moons are generally consistent with his theory. Any deviations were soon explained by and in most cases proven to be due to, the gravity effects of other as yet unknown bodies. His theory helped with the discovery of a number of new planets and explained the strange and long orbit of a comet now called Halley's Comet, much to the delight of Halley himself. He accurately predicted when it would return to been seen from earth. It is unfortunate that last time it came it was so small it was hardly visible. Let's hope it picks up some dust or ice or something for next time it visits us in about another 80 years or so.

However, as new more accurate data became available even Newtons model didn't provide all the answers. As observations improved there were some movements of the solar system that Newton's theory did not fully explain. One of these was the peculiar perihelion advance in the orbit of Mercury. What this means is that the elliptical orbit of mercury is not fixed or aligned in one direction. As it moves around the sun, the closest and furthest points of the orbit slowly change or rotate. This effect was not readily observed or well measured when Newton's laws were being applied to study planetary movement. But subsequent more accurate measurements have clearly determined the magnitude of the effect and it is now estimated to be a significant number of arc seconds per orbit. At one stage this perihelion advance of Mercury's orbit was thought to be due to another planet called Vulcan or some type of dark matter. However no Vulcan or dark matter was ever found. Then a new model of gravity was developed based on G.R. This led to a better planetary model and better orbital calculations. The orbit or Mercury was explainable in terms of this new model and the (light) speed of gravity. This is of course also consistent with TOLG being based on light and acting at light speed although I have yet to finalise the mathematical model and prove it. Parhelion shift is explained below.

ELLIPTICAL ORBIT OF PLANETS
(ACCENTUATED FOR EXPLANATION)

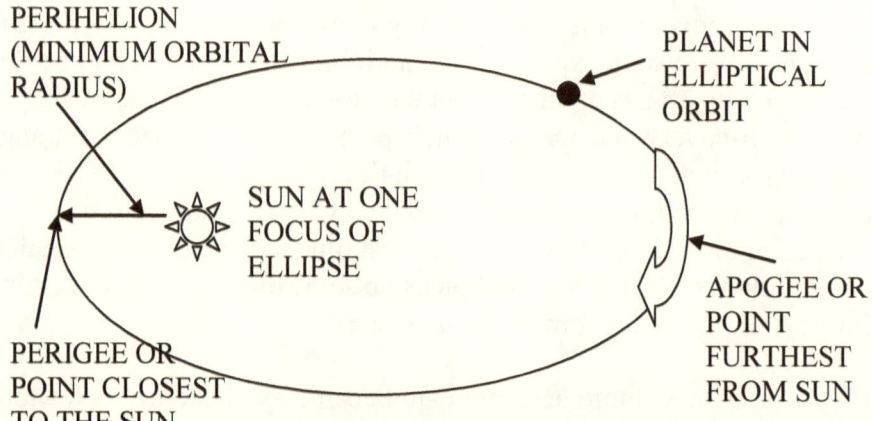

PERIHELION (MINIMUM ORBITAL RADIUS)

PLANET IN ELLIPTICAL ORBIT

SUN AT ONE FOCUS OF ELLIPSE

APOGEE OR POINT FURTHEST FROM SUN

PERIGEE OR POINT CLOSEST TO THE SUN

PEREHELION ADVANCE
(ACCENTUATED FOR EXPLANATION)

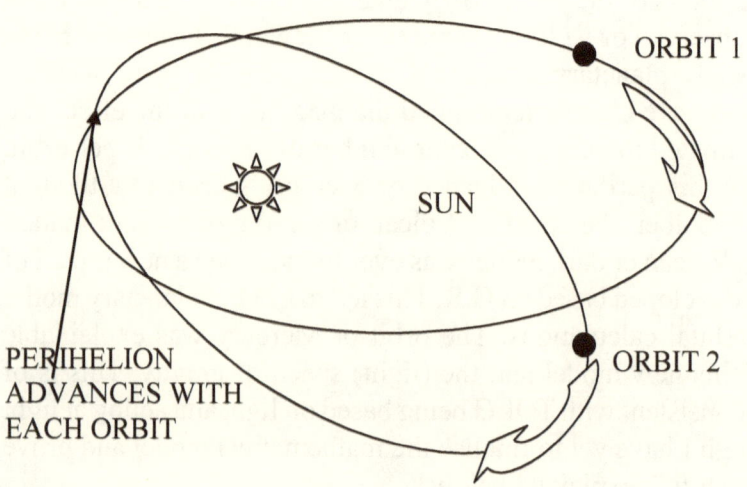

ORBIT 1

SUN

ORBIT 2

PERIHELION ADVANCES WITH EACH ORBIT

While it is most likely that perihelion shift is not peculiar to Mercury, it is more noticeable for Mercury than any other planet due to the fact that Mercury has a more elliptical orbit and a higher orbital speed than most other planets so it has high latency. The same effect should be found for the orbit of Mars as it has a larger latency than Mercury, but because it has a much more circular orbit it may not be easily detectable. Other large outer planets should also show this effect but it would take many years to detect it due to their slow orbits. But it is something for the cosmologists to investigate.

What does TOLG mean to our solar system? To try to answer this question, I examined the solar system in some more detail to see if there were any areas that may be relevant to TOLG. In particular I was looking for latency effects due to the finite speed of gravity and possible shielding or shadow delay effects. The major bodies in the solar system are listed in the attached table along with their key parameters of orbital radius size, mass and speed. One aspect of our solar system that is very relevant to TOLG is the speed of movement of bodies in relevant to background CGBR. TOLG proves that gravity acts at light speed. Therefore any bodies that are moving at high speed (near the speed of light?) will behave differently under TOLG than if they were stationary. This is due to the latency effect where a body is attracted to the previous shadow the attracting body had at a previous time. This will be considerably different from its current position. Does this effect depend on relative speeds between bodies or absolute speeds whatever that means? In any case TOLG will explain perihelion shift and other orbital anomalies.

Another area of interest in gravity calculations is perturbation theory which addresses the "many body problem". This problem is complex due to the summation of the effects of all dynamic forces between many bodies under a gravity model. It is well known that the orbits of all the planets of the solar system suffer perturbation to a greater or lesser degree. In particular the orbits or trajectories of all planets or bodies are a complex function of gravity forces or should I say shadowing with the sun and all other planets. But for bodies nearer Jupiter and especially as they come closer to Jupiter in their orbits the effect will be significant because of Jupiter's massive size. Even the earths orbit is affected by Jupiter as well as by all other planets, especially Venus and Mars. I remember a proposal was once made that earth quakes may be the result of planetary alignment and strong gravitational or perturbation effects. Of course TOLG doesn't change this situation but it may help with calculations associated with it. But what other orbital planetary or man made space trajectory situations may need to be re-examined under TOLG.

PIONEER ANOMALY

The Pioneer Spacecraft launched in the seventies if I remember correctly, has been found to have an unexplained trajectory. For some reason the position is not what is expected from gravity calculations using the current model and it is travelling slower than expected. What is causing the strange trajectory? It has been found to have a deceleration which is not explained, even by G.R. A gas leak was originally thought to be the cause and then ruled out but apparently it has been put back on the agenda. If the deceleration is caused by a leak from on board propellant, how can it last for decades and also how can it be so directed as to only create drag and not just spin. A gas leak for such a long period and with a specific directional effect is unrealistic to say the least. Perhaps the real reason is simply that existing gravity theories are incorrect. What would TOLG do in this situation? Well it may provide the answer because of a mix of factors such as the latency as well as a possible small energy and hence pressure difference due to high speed movement through CGBR. Ouch, red shift due to velocity, can't be! But this may explain the apparent "drag" being observed. Watch this space.

THE SOLAR SYSTEM (NORMALISED TO EARTH DIMENSIONS—EU)

BODY	MASS (EU)	RADIUS (EU)	ORBIT RAD. (EU)	ORBIT TIME (EU).	ORBIT VEL. (EU)	LATENCY DIAMS.	ROTATION (EU)
SUN	330 Thous.	110					29
MERCURY	0.06	0.38	0.39	0.24	1.6	1.90	60
VENUS	0.81	0.95	0.72	0.62	1.17	1.05	243
EARTH	6E+24 Kgm	6378 Km	150 mil. Km	1 year	30 Km/s	1.18	24 hours
MARS	0.11	0.53	1.52	1.88	0.81	2.72	1.02
ASTEROIDS	Small but dangerous	Small but dangerous	2.8	?	?	?	?
JUPITER	310	11.21	5.2	11.86	0.43	0.23	0.41
SATURN	95	9.45	9.5	29.46	0.32	0.38	0.44
URANUS	15	4.01	19	84.00	0.23	1.29	0.72
NEPTUNE	17	3.88	30	165.00	0.18	1.64	0.67
PLUTO **	2.3E-03	0.18	29				
THE MOON	1.2E-04	0.27		28 days			

*Pluto is not considered a planet anymore but is here for reference

CHAPTER 9

TOLG and the Future

So what happens when TOLG is finally accepted? What does it mean for physics in general and in particular the laws of gravity? Will things suddenly fall faster or at different speeds or in different directions? Definitely not! TOLG does not change existing gravity laws in any significant way. Newton's law still applies just as well as it did before. And Einstein's complex mathematics could still be used for most relativistic situations. Nothing will really change in terms of most gravity calculations, at least until the complete mathematical model of TOLG is finalised. Even then the impacts for most people and most situations will be practically zero. However, TOLG will impact on specialist areas of physics and in particular on the Standard Model of cosmology and also on the atomic Standard Model. It should shake some old pillars of physics so to speak but nothing should topple apart from some egos.

Some specific cosmic and atomic impacts are addressed in previous chapters. This chapter reviews some of these but also includes some additional material I had leftover but still wanted to include. Most of this relates to questions about relativity which seems central to everything and to the theories of Special and General relativity. It seems as though the theory of G.R. may be in need of major review, especially the impact on light. In particular if light causes gravity then how can light be affected by gravity. This poses serious problems for Einstein's theory but perhaps there are simple answers to these questions. Let's not throw the bath water out yet just in case there is something in it.

THE BENDING OF LIGHT BY GRAVITY

One of the major outcomes of General Relativity and its approach to gravity was the prediction that gravity bends light. One of the most famous verifications of G.R. was the experiment to measure the bending of light. This was undertaken during an eclipse of the sun shortly after G.R. was published. During an eclipse it was possible to observe the positions of stars near the sun and compare them with their pre and post eclipse positions in cosmic photographs or maps. The results showed a very small deflection of stars near the sun but this was a significant deflection in cosmic terms. It was used to finally "prove" that G.R. was correct. Of course Newton's theory also predicted that gravity will affect a particle with "conventional" mass moving near another body. Newton's calculations produce a solar deflection of about 8 arc seconds assuming that photons have real mass and not just something called rest mass.

Einstein also originally calculated an 8 arc second deflection of light near the sun (as per Newton's theory). But over twice as much deflection was detected (about 18 arc seconds). He eventually doubled his calculations saying that light is bent twice as much. But it is not clear whether this was before or after any experimental measurements. He supposedly explained doubling due to the combined effects of G.R. firstly warping space and then time. But it seems as though such a doubling is not part of any other calculations for normal gravity situations. Perhaps this is because of the different speeds involved. Light is of course travelling at the speed of light and this may cause it to "suffer" double bending. I have not delved into the real background and reasons behind this doubling in G.R. but it would be an interesting review. Of course G.R. experts know the explanation already.

Of course if gravity is caused by light, as proposed by TOLG, then the bending of light by gravity (another type of light) must be seriously questioned. Current light theory doesn't seem to support any such interaction between light beams or photons from entirely unrelated sources. There is no known light or E/M mechanism that would make one unrelated photon cause a change in direction of another unrelated photon. It would seem problematic as both photons have no longitudinal dimensions and perhaps can never "see" each other. What happens to photons in interference situations where correlated ones seem to bounce off each other is another question. So what does this mean to one of the greatest "proofs" of G.R.

Now, I must say that I don't challenge the observation of light bending near the sun that was made by Eddington during an eclipse. However I do challenge its interpretation. Interestingly, recent observations seem to show a more significant bending than that previously observed. Perhaps this is due to improvements in experimental methods and equipment or perhaps it is due to other factors such as changes in the suns atmosphere. Another observation that also seems a bit unusual is that the amount of bending seems to have a large variation depending on the time and place of measurement. That is, when the location of the background starlight being bent is near different parts of the suns surface, while still the same distance from the sun, it is bent by different amounts. Surely this would not be expected if gravity caused it unless the light was actually closer to the sun.

I have another much simpler explanation for the bending of light near the sun. It is based on simple and well established physics for light bending based on the properties of the materials involved. It is well known that the speed of light is not fixed but is determined by two factors, the permittivity ε and permeability (μ) of the region through which it travels (Maxwell). The formula is $c = 1/\sqrt{(\varepsilon\mu}$ Any change in these parameters will change the speed of light. In particular an increase in them for a particular media will result in a slowing of light travelling through that media. In media like water where this effect occurs the speed of light is reduced or light is effectively slowed down. Now as any high school student knows, if light slows down through a media it causes bending. So any increase in these two parameters will cause a slowing and this will obviously cause bending, especially as light passes through the media.

It is also well known that the sun has an "atmosphere" that extends several millions of Km into space. This atmosphere is not a conventional one but is made up of plasma or hot ionised gases. It is also to be expected that hot ionised gases or plasma would have different electric and magnetic properties to free space. In particular plasma would change the permeability and the permittivity of the suns atmosphere relative to free space. It is expected that the suns plasma would increase both of these parameters or at least significantly increase the electric permittivity factor. This increase would result in a (significant) reduction in the speed of light in this region compared to free space and it is this speed reduction that causes bending. So simple! So the "atmosphere" surrounding the sun bends light passing near it. The next question is; by how much? I have yet to estimate the likely change in these parameters and hence the amount of light bending produced but it should be enough to provide a complete explanation for the measured bending? It just surprises me that this plasma bending effect doesn't seem to have been considered in the bending of light by the sun, or at least not in my research on it. Yes I know there have been many other earth bound G.R. gravity light "bending" experiments by well meaning physicists that have further proven the case but I believe that the jury is still out on many of them at best. There may be better reasons for the results of these experiments that have yet to be discovered. Another area for more research?

But what about the other related observations of light bending by gravity such as gravity lensing. Again, it definitely seems to occur and I don't challenge the observation, just the interpretation. Is it possible that the bending of light around other "active" bodies (suns) is also due to an induced speed effect? Can the lensing effect around large cosmic bodies be explained by light speed changes due to plasma or other "gas-atmosphere" effects? I am not sure if this type of effect could cause sufficient bending and hence explain "lensing" in a similar way to how it is explained by G.R., but perhaps it could. Are there any other possible causes such as high level E/M fields that could provide sufficient light bending? Perhaps other possible causes for these observations need to be considered before we automatically "blame" G.R.

ATOMIC FORCES AND TOLG

Does TOLG have any relevance to atomic physics? Well, perhaps not as obviously as for cosmic physics but there are forces at work at atomic level that make the atom what it is. One of these forces is gravity. But what particles do the forces of gravity apply to within subatomic distances? What forces keep electrons and protons "connected" and in mutual orbit in an atomic sense? Both these particles have mass and charge of opposite polarity so it would be expected that both gravity and E/M forces would apply. Or is it simply charged particle attraction or Coulomb forces that keep them in mutual orbits? How does the strong nuclear force apply to this pair of particles? And does gravity keep all nucleons together or is it just protons? In that case what keeps neutrons together? Is it just the weak nuclear force? I am not an expert on this question but the current theory that they are kept together by a combination of nuclear forces including the strong and weak forces may need review. My research will continue but TOLG gravity will surely have a role to play at this level.

A related question is how far into the atom does CGBR penetrate or how small in particle size and mass does CGBR interact? Does TOLG collision occur on the complete nucleus or just protons and hence on what particles does CGBR momentum pressure apply. This question is related to the "size" of CGBR photons (PP) and is discussed in the chapter on light. It is proposed that CGBR impacts on all the larger subatomic particles or nucleons such as protons and neutrons but not electrons. But this area is not well understood, at least by me, and it needs more consideration. Hopefully this will come out of the development of a comprehensive physics analysis of TOLG and a mathematical solution to it and also from work on the "capture" and "training" of PP light.

What role does CGBR play in atomic radio-activity. For example; is PP light in anyway connected with atomic decay. For elements with high atomic number and hence a denser nucleus with more nucleons they are possibly more susceptible to attack and break up from CGBR. There may also be a temporary capture of some small amount of CGBR and not just deflection of it. Is the process of CGBR collision fully non elastic for "heavy" atoms or is there some absorption. Is this process associated with some types of "spontaneous" radioactive decay (fission)? The CGBR gravity theory proposes that most elements do not absorb any energy from CGBR. The reason for this needs more analysis but is obviously related to the frequency and energy of CGBR? That is, electrons can't absorb it and in some way also, nucleons (neutrons/protons) can't (easily) absorb it (when they are in a bound atomic state?). They are just like a mirror which reflects light. A black body absorbs light, so what is a black body for CGBR? This is another area for further analysis and may be a way of examining CGBR using energy absorption theories and possibly high energy research.

SPECIAL RELATIVITY

If gravity is caused by the pressure of CGBR or light as this new theory proposes, then it acts at light speed. This explains a lot about Special

Relativity!! It also shows that nothing can travel faster than light speed at least due to the force of gravity as there can be no gravity pressure (force) faster than CGBR. The other aspect of relativity is that any CGBR shadow that a body creates, changes as the body moves. So if another body feels/sees a CGBR mutual shadow from a moving body, it is in fact the shadow cast by the body when it was in a previous position. The shadow which creates CGBR pressure/gravity is from a previous time and position. As a body moves the mutual shadow it creates changes depending on the speed of the shadow casting body and the distance between the bodies and the speed of light. What this means is that there seems to be an absolute concept of movement or non movement with respect to CGBR. If a body is not moving relative to CGBR then its shadow is also not changing. However if it is moving relatively or is it absolutely relative to CGBR then its shadow is also changing, Of course the movement is related to the relativity between the speed of movement and the speed of light which is very fast but this new concept may have very interesting consequences and may explain some relativistic aspects. What does this mean for red shift? Let me think about that some more.

GENERAL RELATIVITY

The whole theory of General Relativity (G.R.) is of course significantly impacted by this new gravity theory. However, at this stage I have not conducted any detailed research into what aspects of the GR theory may be specifically affected by TOLG (apart from my comments on "gravity" bending light). This is a major area for research. However, understanding the concept of moving shadows due to moving (accelerating?) bodies is probably a key to understanding the relationship between this gravity theory and G.R.

One of the major impacts of TOLG on General Relativity is on the equivalence principle upon which G.R. was supposedly based. TOLG to some extent challenges the correctness of this principle. The principle suggests that accelerating in a gravity field is the same as inertial motion. This is in direct conflict with Newton's law. Equivalence means that it is not possible to tell the difference between the situation when a body has inertial movement in a straight line because no force is making it change (or stay still in a relative sense—whatever this means) and the situation when a force is accelerating a body. The simple example that led Einstein to this idea was of a person falling and accelerating due to some external force such as gravity. The man will "feel" the same as if he was floating freely in free space where no force is being applied. Or the stationary lift with gravity compared to the accelerating rocket with a motor. The occupant can't tell the difference which was interpreted as meaning there is none.

The problem I have always had with this principle is that it uses a human sense or a "feeling" concept. Now we all know that human senses are often deceptive. Not "feeling" a difference is not what I would call a good proof of a scientific theory. If an external force is only applied to one part of your body, such as on your feet, you feel it explicitly. Your bones and muscles and other bodily components then carry the force up and throughout your body to support it and you feel this. You don't feel the downward pressure on all parts of your body due to TOLG. This is a simple earth bound gravity situation. You can feel the (upward) force so to speak even if you are not a Star Wars knight.

But if you are under a gravity force and are in free fall or in orbit around the body creating the force then you will still experience a force on your body, but you may not feel it. Does this really mean there is no force? I don't think so. In TOLG the force is due to a directional difference in PP pressure that applies equally on all atoms (or subatomic particles) in the body. There is no localised pressure or force on any one part of the body such as the feet and therefore there is no transmission of force through the body as in the static earth bound case. This is why you don't feel it. It is still a force but it is unnoticeable. Now to say because you can't feel it means there is no force is stretching the imagination too far. A CGBR pressure "detector" (if and when one is ever built), will be able to detect directional pressure differences and hence a gravity force. But a human won't be able to feel anything. So what does equivalence really mean? Just that you can't feel something. Shades of Ptolemy not being able to feel the earth's very rapid movement.

However, I believe there is a more fundamental problem with equivalence. The two basic situations that are compared are a body not under a force and not accelerating and a body accelerating under a force. The first situation is simply floating in space with no net CGBR pressure difference. There is no force, no acceleration and no energy implications. In the second dynamic situation there are two cases, one is continuous and involves no energy transfer while the second is transient and involves energy transfer. The real issue is; what are the net forces on a body? Is there a net force in one direction? Then what is the direction of movement of the body? Is the movement orthogonal to the direction of the force? For the force to have a direction and for the movement of the body to have a direction, there must be some reference frame for these directions to have meaning. But this is relativity. Perhaps we can use Mach's principle of the universe as a frame of reference. But again let's proceed based on a general understanding of these factors.

When a body experiences a force there are two situations. The first is when the force has a component in the direction of motion. This results in acceleration in the direction of motion. This situation involves energy transfer from the force to the body and therefore due to the laws of energy, cannot last forever. It must therefore be a transient situation and special consideration is required. The only non transient situation is the special case of bouncing through a hole in the body creating the gravity force assuming no friction. This is simple harmonic motion and is similar to circular motion. Energy is conserved and just transfers between kinetic and potential. The second situation is when the force is always orthogonal to the direction of motion of the body and the speed of the body is such as to "balance" this force. This is circular motion and creates an orbiting body which is also simple harmonic motion in a similar way to bouncing through a hole described above. There is no velocity in the direction of the force so no work is done by the force and no energy transfer is involved. It can continue unchanged forever as it has no energy implications again ignoring friction if any. There is angular momentum and of course stored energy as a result of this angular motion just as there is for linear inertial motion. Now is the energy of circular motion, absolute or relative? This is another relativity question shades of Mach's bucket. I am not sure where this leads to but I will continue to think it through. In any case these explanations seem to challenge equivalence.

There are other aspects of G.R. that may need to be reviewed if TOLG is correct, which I believe to be the case. The major ones relate to solutions of G.R. and the so called Big Bang and expanding universe theories. These have been tied to G.R. but Einstein seems to have had major doubts about some of these theories. Perhaps his acceptance of an expanding universe was his "biggest blunder" not the cosmological constant, Λ he inserted to prevent it from happening.

UNSOLVED PP MYSTERIES

Of course as one mystery of the universe is "solved" more questions are raised. In the case of gravity due to CGBR "pressure", the obvious question is; what is the source of CGBR? The answer to this question needs more research but it is most likely to be from some form of atomic activity from very high temperature, high energy cosmic sources way beyond what happens in the sun. Our sun is a good EP generator but perhaps not a very good PP generator if at all. So PP must be very "hot". But it seems that atomic level activity from gravity photons (CGBR) is such that energy is not (easily) transferred to atoms as we know them and gravity photons just bounce around in elastic collisions. Therefore, matter doesn't heat up from the impact of gravity photons; they only impart momentum, not energy!! So CGBR or PP could be said to be cold. That is, the frequency and hence energy level of a gravity photon is so high that it can't be absorbed at atomic (electron) level. PP can't "see" electron clouds and just passes through and bounces of at nucleus level and imparts momentum at nucleus level. That is also the reason why some (most) gravity photons pass through material or have a high degree of penetration with only very little impact. This has implications on the rate of PP required to cause the necessary gravity pressure.

GEOLOGICAL IMPACTS

The earth moves under forces such as gravity and in fact moves in various ways. The two main types of movement are movement as a total rigid body or external movement as I call it, and movement internally which changes its shape or distorts it. External movement is orbiting. The earth's internal movement cause geological forces that create tectonic plate movements or drifts over millions of years. They also make earth quakes and volcanoes and have other effects such as causing high speed Tsunamis. Some of these may be due to internal thermal effects and movement of the core and surrounding molten matter. Some may be due to external gravity pressure differences. There may be CGBR ripple or pressure differences at different times and at different places around the world. The CGBR pressure at each point on earth may vary by direction and over time over and above normal gravity effects from other bodies that keep us in orbit and cause the tides. These may be very small but they may have sufficient impact to change the earths shape slightly and cause earthquakes. This may provide earthquake warnings, once CGBR is found and can be measured.

FURTHER WORK

This new theory is based on sound scientific ideas and methods but just as I have said before it is just a theory. It is open to criticism and review as long as it is scientific and should not simply be rejected by saying "who are you to challenge the greats"? While this new theory of gravity does produce some outcomes that may not yet be readily testable, it is not meta-physics. It should eventually be generally testable, although to prove this new theory may require new technology and new mathematics. There is an urgent need to investigate the new type of light which is the basis for my new theory. The main task will be in finding and identifying this new source of light that creates the force of gravity. But work should start now to contemplate and develop the new field of physics it will lead to it's discovery. Expert research, especially using high energy atom smashing (Hadron accelerating) equipment or a new look at cosmic experiments that are happening all the time, may be useful in this regard. Whatever happens, I am sure this new theory will stand the tests of science and be testable and provable.

As you will have noted during reading this book, I have regularly said this or that area needs more work. Well I was not just saying that to fill the book although that is what appeared to happen. In a way I was asking for help. Perhaps not too subtly but nonetheless I need it. I am not a Professor of physics or even a postgraduate physics student studying for a doctorate (I wish). I am merely a backyard physicist with few ideas. I have very limited resources, not much time and am challenged in the areas of specialist knowledge in physics and mathematics. But I still have a drive to complete this work. I will carry on and hope that others with the required expertise will pick up the challenge and provide input, either negative or positive. All contributions are welcome.

THE END

When I first started to write this book I never thought I would even get this far but here I am at the end of my first ever book. Yes I can hear readers say they never thought they would get this far either. Well let me say thanks for staying the distance. Did you learn anything? Anyway at least I have launched my new theory of gravity, TOLG. Yes, at this stage it is only a basic theory or should I say a proposal as to what causes gravity and it needs more work. In fact it needs lots of work to finalise and prove it and then make it useful and possibly derive benefits from it. This is all perhaps still a long way off.

While there may be no major changes to gravity calculations due to TOLG there should be implications on most areas of physics, more than have been identified in this brief study. So let the study begin to find and evaluate them and move forward in new directions that are created as a result. Physics hasn't had such a shake up since Einstein was a lad.

But before you put the book down just a word about my next book. This time it is on the related subject of light. I want to find the real cause of red shift and interference and many other properties of light. I have already developed many ideas but this book was getting too big so I didn't include many of them. After that perhaps I will leave physics for a while and write on a completely different subject, that of human civilisation, evolution of ideas, global economics and how to handle the future we are making for ourselves or at least our kin. I haven't written it yet so you may have to wait a short while but again the wait will be worth it.

www.ingramcontent.com/pod-product-compliance
Lightning Source LLC
Chambersburg PA
CBHW032002170526
45157CB00002B/504